Military Applications of Data Analytics

Data Analytics Applications

Series Editor: Jay Liebowitz

PUBLISHED

Actionable Intelligence for Healthcare
Jay Liebowitz and Amanda Dawson

Data Analytics Applications in Latin America and Emerging Economies
by Eduardo Rodriguez

Big Data and Analytics Applications in Government: Current Practices and Future Opportunities
by Gregory Richards

Big Data Analytics in Cybersecurity
by Onur Savas and Julia Deng

Data Analytics Applications in Education
by Jan Vanthienen and Kristoff De Witte

Intuition, Trust, and Analytics
by Jay Liebowitz, Joanna Paliszkiewicz, and Jerzy Gołuchowski

Research Analytics: Boosting University Productivity and Competitiveness through Scientometrics
by Francisco J. Cantú-Ortiz

Big Data in the Arts and Humanities: Theory and Practice
by Giovanni Schiuma and Daniela Carlucci

Analytics and Knowledge Management
by Suliman Hawamdeh and Hsia-Ching Chang

Data-Driven Law: Data Analytics and the New Legal Services
by Edward J. Walters

Military Applications of Data Analytics
by Kevin Huggins

For more information about this series, please visit:
https://www.crcpress.com/Data-Analytics-Applications/book-series/CRCDATANAAPP

Military Applications of Data Analytics

Edited by

Kevin Huggins

CRC Press is an imprint of the
Taylor & Francis Group, an **informa** business
AN AUERBACH BOOK

CRC Press
Taylor & Francis Group
6000 Broken Sound Parkway NW, Suite 300
Boca Raton, FL 33487-2742

First issued in paperback 2022

ISBN 13: 978-1-03-247601-8 (pbk)
ISBN 13: 978-1-4987-9976-8 (hbk)

DOI: 10.1201/9780429445491

Visit the Taylor & Francis Web site at
http://www.taylorandfrancis.com

and the CRC Press Web site at
http://www.crcpress.com

Contents

Preface

Military organizations around the world are normally huge producers and consumers of data. Accordingly, they stand to gain from the many benefits associated with data analytics. However, for leaders in defense organizations—either government or industry—accessible use cases are not always available. The reason that I embarked on this book was to provide a diverse collection of detailed use cases that could serve as a representation of the realm of possibilities in military data analytics. I trust you will find it useful, and I look forward to hearing about your successful analytics projects.

Acknowledgments

This, like most other works, was truly a team effort. First, I'd like to thank my family for their patience and encouragement. Additionally, I'm grateful to Jay Liebowitz for giving me this opportunity. It was definitely an adventure. I am also indebted to my colleagues at Harrisburg University for their insights and recommendations. Finally, I would like to thank my awesome team of reviewers. They generously shared their expertise to raise the quality of this work significantly. The review team consisted of the following experts:

- Nathaniel Ashby, PhD
- Feyzi Bagirov
- Rand Ford, PhD
- Philip Grim
- Marvine Hamner, PhD
- Pablo Ibargüengoytia, PhD
- Suzanne J. Matthews, PhD
- Arnie Miles
- Glenn Mitchell, MD
- Ali Motamedi, PhD
- Stephen Penn, PhD
- Kevin Purcell, PhD
- Mandana Rezaeiahari, PhD
- Roozbeh Sadeghian, PhD
- Aaron St. Leger, PhD
- Doaa Taha, PhD

Editor

Kevin Huggins, PhD, is professor of Computer Science and Analytics at Harrisburg University of Science and Technology, Harrisburg, Pennsylvania. He is also a retired military officer who spent the early part of his career in military intelligence, with extensive experience in Latin America. The remainder of his career was dedicated to academia, primarily as a faculty member in the Department of Electrical Engineering and Computer Science at the U.S. Military Academy. While there, Dr. Huggins served as the director of Research in Network Science as well as the director of the Information Technology Program.

Additionally, Dr. Huggins was a visiting scientist at the École de Techniques Avancées in Paris, France, where he studied parallel algorithms for multiprocessor system-on-chip (MPSoC) architectures. His current research interest lies at the intersection of data science and information security, exploring novel ways of securing computing systems by leveraging the enormous amounts of available data.

He holds a PhD in computer science from Mines Paris Tech.

Contributors

David Aebischer is chief of the Special Operations Branch of the U.S. Army Communications Electronics Command (CECOM) Training Support Division (TSD) and is the inventor and leader of the CECOM Equipment Diagnostic Analysis Tool, Virtual LAR (CEDAT VLAR) project, which uses advanced operations research techniques to develop diagnostic applications for Army combat equipment. Aebischer is a veteran of the U.S. Air Force with over 36 years of government service—including over 4 years deployed to Iraq and Afghanistan supporting soldiers and their equipment. Aebischer holds a master's of business administration and technology management degree from Monmouth University and is pursuing a doctoral degree in applied management and decision science. Aebischer is a class of 2016 Franz Edelman laureate and his VLAR team was selected as a finalist for the 2016 Franz Edelman award for significant contribution to operations research and innovation in applied analytics. He has extensive technical training and field experience in electrical, electromechanical, and electronics systems and has advanced training on Bayesian networks and Cognitive Task Analysis (CTA) processes. Aebischer's current research interest is in using Bayesian networks to optimize fuel consumption and equipment readiness.

Chris Arney is a professor of mathematics at the U.S. Military Academy. He helped to develop a network science minor at the academy and often teaches courses in modeling, interdisciplinary problem solving, and differential and difference equations. He served in the active duty Army for 30 years, was a dean and vice president at a college, and directed and managed research programs in network science, mathematics, and cooperative systems. He is the founding director of the Interdisciplinary Contest in Modeling.

Karna Bryan has worked for over 20 years on data-driven decision-making research topics in various domains including underwater warfare, maritime domain awareness, and civil aviation. She currently works as an operations research analyst for the Department of Transportation in Cambridge, Massachusetts. Previously, she worked for 18 years at the Center for Maritime Research and Experimentation and was program manager for the Maritime Security program. She holds a master's degree in statistics from Yale University.

Susan Dass is a senior instructional designer and technical specialist at ICF. She is also an adjunct professor at George Mason University. Dass received a PhD in education and an MEd in instructional design. She has more than 17 years of experience in instructional design supporting both private and government entities. Dass has written many publications, including coauthoring three book chapters that explore the design of help-seeking and self-regulated learning skills in 3D learning environments as well as faculty adoption of virtual worlds as a learning environment. Her current projects include designing 3D serious games for the medical industry and exploring data analytics for learner performance and course improvements.

Clarence Dillon is a senior consultant with ICF and has been working with data, strategic planning, and management for 24 years, including time working with the Office of the Secretary of Defense, the Office of the Secretary of the Army, the Army staff, and the Air Force staff. He is a computational social science PhD candidate in the Department of Computational and Data Sciences at George Mason University. He holds an MSocSci from the University of Tampere, Finland, and a BA from the University of Colorado at Boulder.

Uriel A. García received a BsSc degree in engineering in computer systems from the Technological Institute of Comitan, Chiapas, Mexico. Currently, he is pursuing a master's degree in optimization and applied computing at the Autonomous University of Morelos, Mexico. His research interests are focused on machine learning, artificial intelligence in renewable energy, and probabilistic graphical models.

José Alberto Hernández-Aguilar obtained a PhD from Autonomous University of Morelos (UAEM) in 2008, a master's degree (summa cum laude) in business management (2003) from the Universidad de las Américas Puebla (UDLAP), and a BSc in computer engineering from the National University of Mexico (UNAM). He has been a full-time professor since 2010, at the Accounting, Management and Computer Sciences School at UAEM. His main interest areas are databases, artificial intelligence, online assessment systems, data mining, and data analytics. He is a member of the organizing staff of different international congresses such as the Healthcare Infection Society (HIS) and the Mexican International Conference on Artificial Intelligence (MICAI).

Javier Herrera-Vega received BSc and MSc degrees in computer science from the Autonomous University of Puebla, Mexico. Currently, he is finishing a PhD degree in computer science from the National Institute of Astrophysics, Optics and Electronics in Puebla, Mexico. His research interests are focused on inverse problems, probabilistic graphical models, and medical imaging.

Ziyuan Huang is an experienced SQL server developer with experience in the financial services and healthcare industries. He also has experience in analytical skills, SQL Server R services, SQL Server Reporting Services (SSRS), Extract, Transform, Load (ETL), databases, and Microsoft Access. Huang is currently a PhD candidate in data sciences at Harrisburg University of Science and Technology.

Pablo Ibargüengoytia received a degree in electronic engineering from the Universidad Autónoma Metropolitana (UAM). He received a master's degree from the University of Minnesota. Finally, he received a PhD in computer science from the University of Salford in the United Kingdom. He has been a full-time investigator at the National Institute of Electricity and Clean Energies (INEEL) since 1983. He has been a professor of intelligent systems at the ITESM Campus Cuernavaca since 1998. He has also directed several BSc, MSc, and PhD theses. He is a senior member of the Institute of Electrical and Electronics Engineers (IEEE), a member of the Mexican Society of Artificial Intelligence, and a member of the National Researchers System (SNI) in Mexico.

Anne-Laure Jousselme is with the NATO Centre for Maritime Research and Experimentation (CMRE) in La Spezia, Italy, where she conducts research activities on maritime anomaly detection, high-level and hard and soft information fusion, reasoning under uncertainty, information quality assessment, and serious gaming approaches. She is a member of the board of directors of the International Society of Information Fusion (ISIF) where she serves as vice president of membership. She serves on the board of directors and steering committee of the Belief Functions and Applications Society (BFAS) as well as on program committees of the FUSION and BELIEF conferences. She co-organized the Information Fusion conference in Quebec City in 2007 (tutorial chair) and served on the organizing committee of Fusion 2015 in Washington (international co chair). She is associate editor of the *Perspectives on Information Fusion* magazine. She has been an adjunct professor at Laval University, Quebec, Canada for 10 years.

Rodney Long is a science and technology manager at the Army Research Laboratory in Florida and is currently conducting research in adaptive training technologies. Long has a wide range of simulation and training experience spanning 29 years in the Department of Defense and has a bachelor's degree in computer engineering from the University of South Carolina, as well as a master's degree in industrial engineering from the University of Central Florida.

Suzanne J. Matthews is an associate professor of computer science in the Department of Electrical Engineering and Computer Science at the U.S. Military Academy. She received her BS and MS degrees in computer science from Rensselaer Polytechnic Institute and her PhD in computer science from Texas A&M University. Her honors include a dissertation fellowship from Texas A&M University, a master teaching fellowship from Rensselaer Polytechnic Institute, and memberships in the honor societies of Phi Kappa Phi and Upsilon Pi Epsilon. She is a member of the Association of Computing Machinery, and an affiliate of the IEEE Computer Society. Her research interests include parallel computing, single board computers, data analytics, experimental computing, and computational biology.

Eduardo F. Morales received a PhD from the Turing Institute at the University of Strathclyde, in Scotland. He has been responsible of more than 25 research projects and has written more than 150 peer-review papers. He was an invited researcher at the Electric Power Research Institute (1986), a technical consultant (1989–1990) at the Turing Institute, a researcher at the Instituto de Investigaciones Electricas (1986–1988 and 1992–1994), and at ITESM–Campus Cuernavaca (1994–2005). He is currently a senior researcher at the Instituto Nacional de Astrofísica, Óptica y Electrónica (INAOE) in Mexico where he conducts research in machine learning and robotics.

Mark Newman is an experienced senior engineer with a demonstrated history of working in a various data-heavy industries. Skilled in Azure, .Net, Microsoft SQL, and Microsoft Dynamics CRM, Newman is a strong engineering professional with an MS focused in computer science from Hood College. He is currently a PhD candidate in data sciences at Harrisburg University of Science and Technology.

Norma Josefina Ontiveros-Hernández holds an MSc in computer science (1995) from the Centro Nacional de Investigación y Desarrollo Tecnológico (CENIDET), Mexico (software engineering) and a BSc in computer science from Durango Institute of Technology (1984). She has lectured at the Mexico National System of Technology since 1986, at Durango Institute of Technology from 1986 to 1997, and at Zacatepec Institute of Technology since 1997. Her primary areas of interest are databases, artificial intelligence, assessment systems, and analytics.

Miguel Pérez-Ramírez holds a PhD in computer science (artificial intelligence from the University of Essex, UK; MSc in computer science (software engineering) from Centro Nacional de Investigación y Desarrollo Tecnológico (CENIDET), Mexico, and a BSc in computer science from Benemérita Universidad Autónoma de Puebla (BUAP), Mexico. He has been a research member of the Mexican National Institute for Electricity and Clean Energies, since 1992, in the information technologies department. He has participated in different projects, which include development of information systems and data warehouses. He has also worked with expert systems and knowledge management. Since its creation in 2003, he has led the Virtual Reality Group, which focuses on the development of training systems for the energy sector, where the learning context is high-risk tasks, such as maintenance to energized lines within the electrical domain. These training systems have been installed and used across the entire country in the Comisión Federal de Electricidad (CFE), the Mexican electricity utility company. He is a member of the IEEE.

Kevin Purcell is the chief scientist at WildFig, a data science and analytics consultancy. Purcell has an MS and PhD in the life sciences and conducts research focused on applying computational and statistical methods to understand disturbance in complex systems. He is currently an adjunct faculty member in the Analytics Department at Harrisburg University, teaching graduate courses in data science. He currently heads WildFig's first research and experiential learning laboratory, with a mission to develop novel analytical approaches with commercial applications and cultivate a new generation of data scientists.

Sridhar Reddy Ravula is an entrepreneur, asset manager, and business consultant. He is passionate about solving business problems and providing cost-effective solutions using emerging technologies. He holds an MBA, with specializations in finance and operations, and an MS in data analytics. He is currently pursuing a PhD in data science at Harrisburg University. His research interests are applications of natural language processing and AI in the domains of cybersecurity, finance, and healthcare.

Joshua Rykowski is the director for Strategic Initiatives for the Cyber Protection Brigade at the U.S. Army Cyber Command.

Michael Smith is a senior technical specialist with ICF and has over 14 years of experience in data analytics, strategic planning, and management. Smith advises government clients on how to adapt analytics practices to improve organizational performance. He has a BA in international economics from Longwood University and a master's degree in public policy from Georgetown University.

Matthew Sobiesk has a BA in mathematics and Near Eastern studies from Cornell University and is currently an operations research PhD student at the Massachusetts Institute of Technology. His research interests include studying machine learning through an optimization lens.

Aaron St. Leger is an associate professor at the U.S. Military Academy (USMA). He received BSEE, MSEE, and PhD degrees from Drexel University. His research and teaching interests include alternative energy, electric power systems, modeling, and controls. He has published more than 50 papers on these subjects. His recent work focuses on integrating alternative energy and demand response controllers to improve electric power systems for military forward operating bases, and anomaly detection in smart grids. He is the director of the Electrical Power Systems, Alternative Energy, and Operational Energy Laboratories at USMA.

Ralph O. Stoffler is a member of the Senior Executive Service, is the director of Weather, deputy chief of staff for Operations, Headquarters, U.S. Air Force, Washington, DC. In this capacity, he is responsible for the development of weather and space environmental doctrine, policies, plans, programs, and standards in support of Army and Air Force operations. He oversees and advocates for Air Force weather resources and monitors the execution of the $320 million per year weather program and is the functional manager for 4300 personnel.

L. Enrique Sucar has a PhD in computing from Imperial College, London, an MSc in electrical engineering from Stanford University, and a BSc in electronics and communications engineering from ITESM, Mexico. Currently, he is a senior research scientist at the National Institute for Astrophysics, Optics and Electronics, Puebla, Mexico. Sucar is a member of the National Research System and the Mexican Science Academy. He is the associate editor of the *Pattern Recognition and Computational Intelligence* journal, and a member of the advisory board of International Joint Conferences on Artificial Intelligence (IJCAI). He was awarded the National Science Prize by the Mexican president in 2016.

Natalie Vanatta is a U.S. Army cyber officer and currently serves as the deputy chief of research at the Army Cyber Institute. Here, she focuses on bringing private industry, academia, and government agencies together to explore and solve cyber challenges facing the U.S. Army in the next 3 to 10 years in order to prevent strategic surprise. She holds a PhD in applied mathematics, as well as degrees in computer engineering and systems engineering. Vanatta has also served as distinguished visiting professor at the National Security Agency.

Benjamin Eddie Zayas-Pérez is a researcher at the Supervision Department of Electrical Research Institute. He obtained a PhD in computer sciences and artificial intelligence from Sussex University, following predoctoral studies in human-centered computer systems and computer sciences. His research interests include human–computer interaction and virtual reality environments for learning and training purposes. Currently, he is initiating applied research in Big Data analytics. He is member of the IEEE.

Chapter 1

Bayesian Networks for Descriptive Analytics in Military Equipment Applications

David Aebischer

Contents

Introduction

The lives of U.S. soldiers in combat depend on complex weapon systems and advanced technologies. Operational command, control, communications, computers, intelligence, surveillance, and reconnaissance (C4ISR) systems provide soldiers the tools to conduct operations against the enemy and to maintain life-support. When equipment fails, lives are in danger; therefore, fast and accurate diagnosis and repair of equipment may mean the difference between life and death. But in combat conditions, the resources available to support maintenance of these systems are minimal. Following a critical system failure, technical support personnel may take days to arrive via helicopter or ground convoy, leaving soldiers and civilian experts exposed to battlefield risks. What is needed is a means to translate experiential knowledge and scientific theory—the collective knowledge base—into a fingertip-accessible, artificial intelligence application for soldiers. To meet this need, we suggest an operations research (OR) approach to codifying expert knowledge about Army equipment and applying that knowledge to troubleshooting equipment in combat situations. We infuse a classic knowledge-management spiral with OR techniques: from socializing advanced technical concepts and eliciting tacit knowledge, to encoding expert knowledge in Bayesian Belief Networks, to creating an intuitive, instructive, and learning interface, and finally, to a soldier internalizing a practical tool in daily work. We start development from this concept with counterfactuals: What would have happened differently had a soldier been able to repair a piece of critical equipment? Could the ability to have made a critical diagnosis and repair prevented an injury or death? The development process, then, takes on all the rigor of a scientific experiment and is developed assuming the most extreme combat conditions. It is assumed that our working model is in the hands of a soldier who is engaged with the enemy in extreme environmental conditions, with limited knowledge of, and experience with, the equipment that has failed, with limited tools, but with fellow soldiers depending on him (her) to take the necessary steps to bring their life-saving equipment back into operation and get the unit out of danger (Aebischer et al., 2017). The end product is a true expert system for soldiers. Such systems will improve readiness and availability of equipment while generating a sustainable cost-savings model through personnel and direct labor reductions. The system ensures fully functional operation in a disconnected environment, making it impervious to cyber-attack and security risks. But most importantly, these systems are a means to mitigate combat risk. Reducing requirements for technical support personnel reduces requirements for helicopter and ground-convoy movements, and this translates directly to reductions in combat casualties (Bilmes, 2013).

In *On War*, Carl Von Clausewitz wrote, "War is the realm of uncertainty; three quarters of the factors on which action is based are wrapped in a fog of greater or lesser uncertainty" (Kiesling, 2001). Clausewitz's "Fog of War" metaphor is well-made and oft-used, becoming so popular as to become part of the modern military lexicon. Analysis of his writing has, naturally, led to discussion about how information technology can be brought to bear to reduce battlefield entropy. From this familiar territory, let us deliberately examine data analytics in context with combat uncertainty. Let us further refine that to the combat soldier and his equipment—in the most extreme scenario of enemy threat, geography, and environment—and determine what data analytics can put at a soldier's fingertips. This task requires a fine balance of complexity and usability. It must be mathematically precise, yet infused with tacit knowledge.

It must be exhaustive, as the consequences of failure can be dire. Most of all, the task must establish the primacy of soldiers at the point of need. We submit that this is a doable and necessary task and an opportunity to apply a unique analytics solution to this, and other, multivariate problems. This chapter will provide an introductory guide for tapping into an expert knowledge base and codifying that knowledge into a practical, useful, and user-friendly working model for diagnostics. In order, we will provide a general description of all the tools, techniques, and resources needed to construct the model, and then proceed through a practical example of model building.

The category of Descriptive Analytics, and diagnostics as a sub-category, serves well to frame our initial efforts. Korb and Nicholson (2011) describe:

> By building artifacts which model our best understanding of how humans do things (which can be called descriptive artificial intelligence) and also building artifacts which model our best understanding of what is optimal in these activities (normative artificial intelligence), we can further our understanding of the nature of intelligence and also produce some very useful tools for science, government, and industry. (p. 21)

Causal analysis is at the heart of descriptive analytics. To perform this analysis, we will need a couple of tools: a knowledge engineering tool to capture knowledge about the domain of interest, and Bayesian networks to codify and represent that knowledge. The knowledge engineering tool and the Bayesian network tool work hand-in-hand, so they will be detailed as such. We will also need to designate a specific C4ISR system as a use case and as a practical example to demonstrate how each tool works separately and in combination with the other to model complex systems. For this, we will use the Army's diesel engine-driven tactical generator, the primary source of power for Army Command Post operations. Generators meet our needs in terms of multi-domain complexity (electrical, electronics, and electro-mechanical) and in terms of ubiquity in combat environments. We will look at these tools individually in some detail, but we will focus more on how they overlap and how they integrate around our diesel generator use case. All of this goes toward learning how to work with limited or nonexistent data. Our richest source of knowledge is experts, and our actual data-generating process is from facilitating effective and efficient knowledge elicitation sessions with those experts. Our first tool will help us navigate the challenging, but rewarding, world of working with experts and expertise.

It is easy to recall instances where we have witnessed experts at work. The great artists—composers, painters, poets, musicians—seem to possess something that transcends talent, and they describe their art as something that occurs when their ideas have gone somewhere that was not intended. The process by which they create and do is not describable, existing only in their heads. It is much the same with equipment experts—possessing a blend of experiential and theoretical knowledge and an innate ability to apply that knowledge in ways few of us can comprehend. Coming to understand, and make explicit, even a fraction of this knowledge is the goal of knowledge engineering and the process of knowledge elicitation (Figure 1.1).

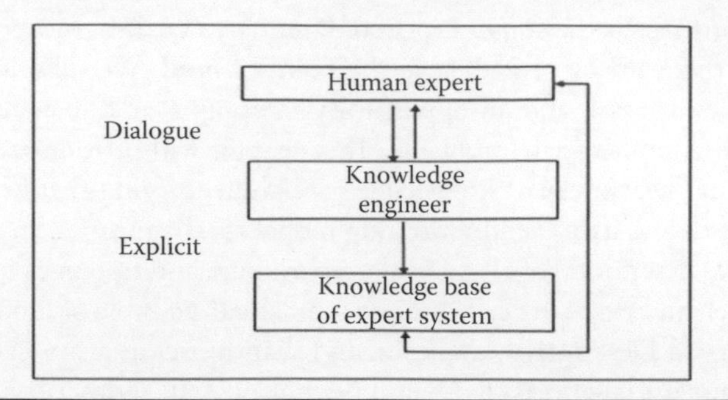

Figure 1.1 Knowledge engineering.

This process is painstaking, labor-intensive, and fraught with the danger of introducing many different biases, but the benefits of a well-constructed and executed knowledge elicitation make it worth the effort and outweigh the risks. It is in this process that we tease out the heuristic artifacts buried in the brains of experts. It is these artifacts that are carefully crafted into the artificial intelligence network that sits in a soldier's hands.

The knowledge engineering process has its roots firmly in Cognitive Task Analysis (CTA). CTA involves a number of processes by which decision making, problem solving, memory, and judgment are all analyzed to determine how they are related in terms of understanding tasks that involve significant cognitive activity. For our purposes, CTA supports the analyses of how experts solve problems.

DSEV

Define, Structure, Verify, Elicit (DSEV) is a field-proven method for developing Bayesian models strictly from expert knowledge with a unique blend of soft and hard skills. DSEV is infused with techniques that mitigate bias and ensure the elicitation process aggregates expert judgments into a unified network. We will first outline the methods we use to execute each phase of the model and then, later, demonstrate how it is applicable, flexible, and scalable to a practical problem domain.

DSEV starts, stops, and cycles with experts. In both the theory and practice of Bayesian networks, assumptions—expert knowledge about any specific problem domain—are a necessary component. But it is inherently difficult for experts to explain what they know, and equally difficult for non-experts to understand what experts are saying. What is needed is a formalism for bridging the gap between experts and non-experts and codifying complex technical concepts into graphical structure and probability distributions. Our DSEV objective is an exhaustive probabilistic model of how the generator can fail—with expert-based pathways through each failure mode. But DSEV requires unique facilitation skills, so, prior to starting the DSEV process, we want to do some preparatory tasks so that facilitators and experts—the whole team—understand the tools, tasks, and processes associated with Figure 1.2 (Hepler et al., 2015).

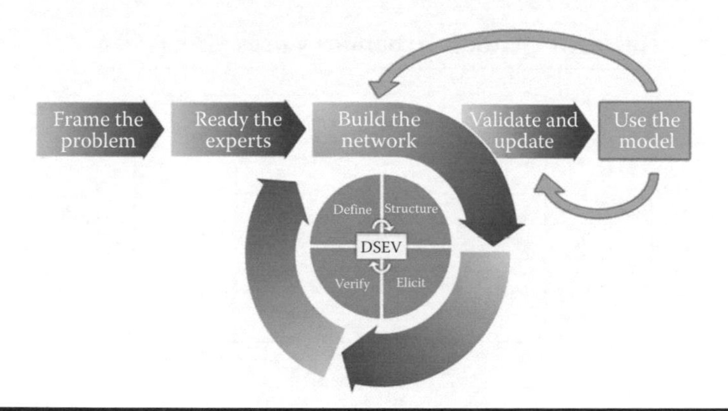

Figure 1.2 Define, Structure, Elicit, Verify (DSEV).

Frame the Problem

This part of the DSEV includes establishing and scoping objectives, identifying use cases, and conducting preliminary research. While our overarching objective is an exhaustive probabilistic model of generator failure modes, we need to properly scope that objective for a discrete DSEV session. In practice, this means breaking the generator system into logical subsystems and working on modeling each of those subsystems in turn. Defining use cases then becomes a straightforward task related to the subsystem: operation of the system in normal conditions, operation under abnormal/extreme environmental conditions, and operation under other special conditions specific to the combat scenario. The last step in this pre-phase block of the process is to ensure everybody on the team is speaking the same technical language. It is incumbent upon the facilitators to be conversant with the subject matter (theory, operation, technical terms, acronyms) and to know what questions will be most effective. Background research, training classes, and seminars are all ways for facilitators to ready themselves to talk to experts. CTA techniques, such as the critical decision method (CDM), provide a framework for how to structure the conversation. We will delve into details of CDM during the practical modeling portion of the chapter. With objectives and use cases agreed to and a well-versed facilitator team in place, it's time to organize the group.

Ready the Experts

Deciding how many, what types, and what combination of experts to use in an elicitation is the first challenge. Statistically, four to five experts is the right amount to mitigate some biases and to avoid bogging the discussion down with *too* many points of view and *too* much interaction between members. This is directly related to the number of possible social interactions in the group dynamic. Above a group of five, this number grows exponentially (Table 1.1).

Table 1.1 Group Dynamics versus Group Size

Research on practical group events bears that out, with group member ratings converging to between four and five members for optimum efficiency (Hackman and Vidmar, 1970). It is usually best to choose "complementary experts": some with specific hands-on experience with the equipment, and some with more academic or theoretical perspectives. It also helps to have an expert with significant general knowledge in the field (domain experience) but not specific experience on the system. This combination provides a good setting for "productive friction" between experts, stimulating the CDM "deepening" process that reveals new insights into the domain. The expert team is combined with a facilitation team consisting, normally, of two knowledge engineers, with one focusing on doing the direct elicitation—extracting and interpreting the expert knowledge—and one focusing on building a model based on the elicitation and interpretation. Regular dialogue between knowledge engineers to clarify statements is expected, and it is often helpful to reverse roles at some interval during the elicitation. With a well-organized and well-balanced team in place, the next step is to ensure the experts know how their knowledge will be used and how will it be represented.

Bayesian networks (BNs) are probabilistic models of how we believe our "world" works (Pearl, 1988). BNs are based on Bayes theorem, an equation that calculates the probability of an event (A) given we know an event (B) has occurred, or P (A|B) (Figure 1.3).

Bayes Theorem

- Bayes theorem is most commonly used to estimate the state of a hidden, causal variable ***H*** based on the measured state of an observable variable ***D***:

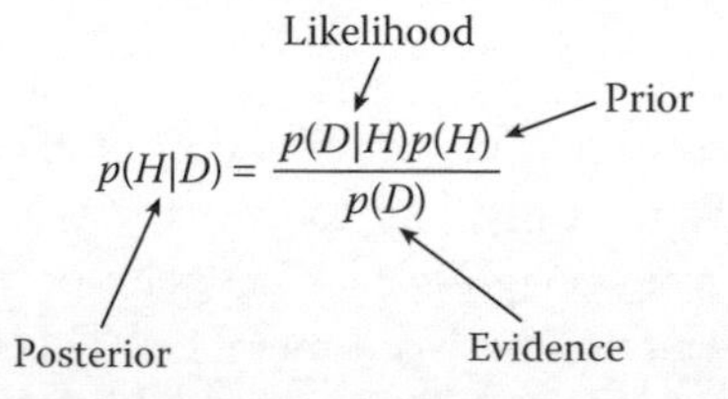

Figure 1.3 Bayes theorem.

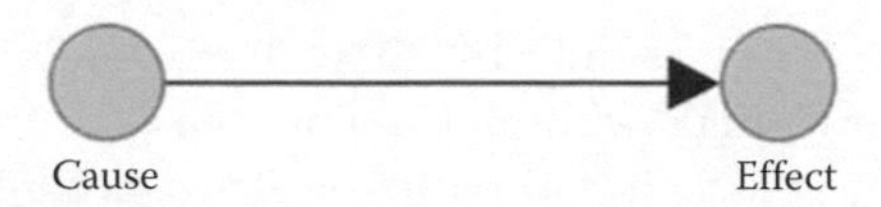

Figure 1.4 Causal relationships in a Directed Acyclic Graph (DAG).

BNs have a qualitative component and a quantitative component. The qualitative, or structural, component is a Directed Acyclic Graph (DAG). DAGs are constructed using nodes, representing variables in the domain of interest, and arcs, representing the causal (temporal) relationship between variables (Figure 1.4).

Relationships between variables can be modeled using any of three basic structures, individually or in combination: (1) indirect connection (A causes B, which causes C); (2) common cause (A can cause B and C); or (3) common effect (A or B can cause C). Each structure has unique characteristics that guide the process of choosing how to represent specific relationships (Figure 1.5).

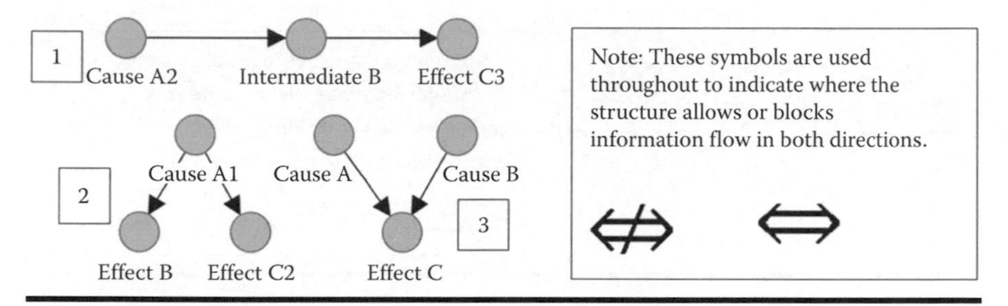

Figure 1.5 Basic structures.

In an indirect connection structure, A and C are marginally dependent (each variable is in its natural distribution with no evidence entered). For example, knowing that the ambient temperature is cold changes our belief about an engine starting failure through an open path between A (Ambient Temperature) and C (Engine Cranks but Will not Start) through B (Battery) (Figure 1.6).

Critical to the understanding of Bayesian networks is the concept of omnidirectional inference: the path is open in both directions. In the small network

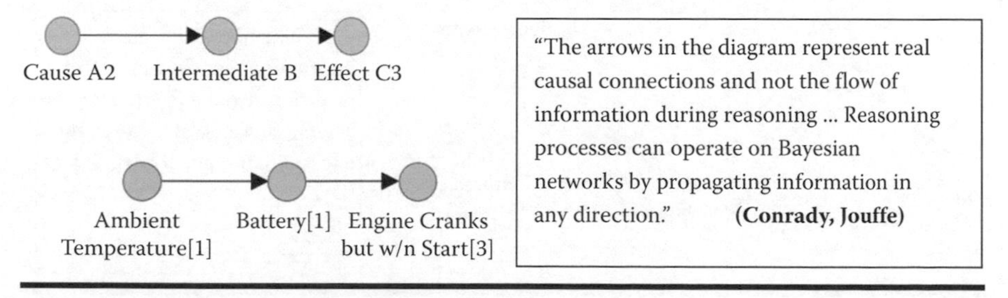

Figure 1.6 Indirect connection structure.

above, the arcs do indicate a cause and effect, or temporal, ordering. But information flows in both directions—both with and against the arc direction. So knowing the Engine Cranks but w/n Start and setting evidence accordingly will influence the probability distribution of Battery and Ambient Temperature, just as setting evidence on Ambient Temperature will influence the distribution of Battery and Engine Cranks but w/n Start (Figure 1.6).

However, if we know B (Battery state of charge is **LOW**, as in Figure 1.7), and set evidence accordingly, this closes the information path in both directions. We say, then, that A and C are independent given (conditioned on) B, or that setting any evidence on B makes A and C independent. This is intuitive because if we know the Battery state of charge is low or high, then knowing Ambient Temperature is no longer useful with respect to the "Engine Cranks but w/n Start" variable, and vice versa. This structure type is particularly useful for modeling electrical wiring or fiber optic runs, as it will translate an expert's thought such as "a reading at test point B will make a test point A reading unnecessary with respect to C" into a compact structural model.

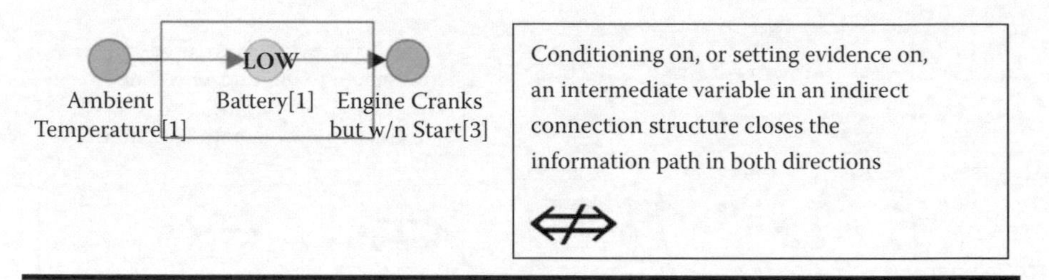

Figure 1.7 Indirect connection—Evidence.

In the common cause structure (A can cause B or C), B and C are marginally dependent. Information can flow from B to C through A, and from C to B through A. Consider a case where the "Cylinder Pressure" variable (A) can be the cause of both the "Engine Cranks but w/n Start" (B) and "Compression Test" (C) variables. Knowing the state of the "Engine Cranks but w/n Start" variable will certainly change our reasoning (and probability distribution) on the "Compression Test" variable, and vice versa (Figure 1.8).

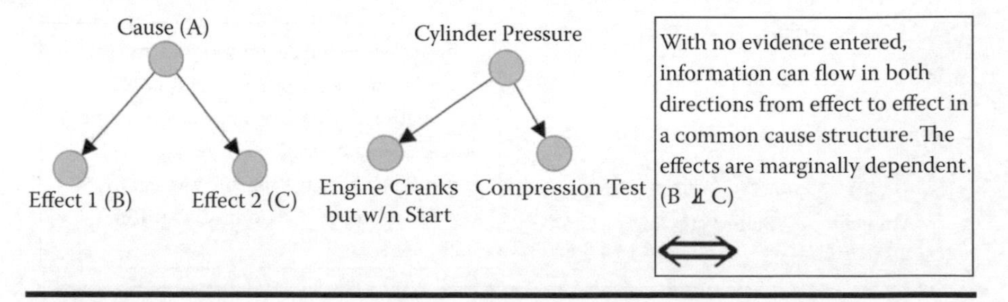

Figure 1.8 Common cause (confounder) structure.

However, if we already know the state of the "Cylinder Pressure" variable (we know the cause) and enter evidence accordingly, this blocks the path between effects through the cause in both directions. This is intuitive because knowing the state of "Cylinder Pressure" makes knowing the state of "Engine Cranks but w/n Start" no longer useful to our belief about "Compression Test," and vice versa. B and C are, then, independent given evidence for A (Figure 1.9).

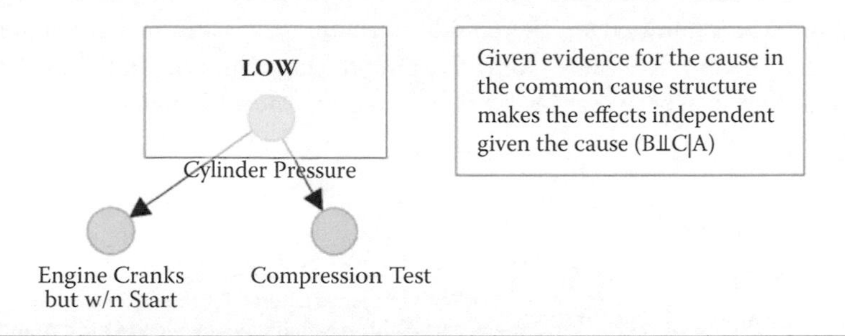

Figure 1.9 Common cause (confounder)—Evidence.

The common cause structure translates expert's statements such as "low cylinder pressure can cause engine starting problems and a marginal or low compression test" directly into a compact model that reasons like the expert; that is, what tests are most economical and most sensitive?

In the common effect structure, A or B can cause C. Comment effect structures are also known as colliders. Information cannot flow from A to B in a collider because C blocks the path of information in both directions. A and B are, then, marginally independent. Consider a case where both the "Battery" and the "Fuel System" could cause "Engine Cranks but w/n Start." This is intuitive, as knowing the state of "Battery" tells us nothing about "Fuel System" (Figure 1.10).

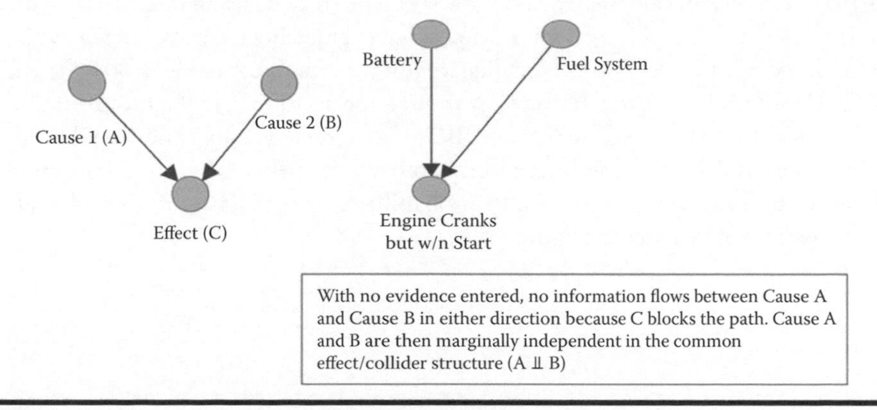

Figure 1.10 Common effect (collider) structure.

But, if we know the state of "Engine Cranks but w/n Start" and we enter evidence accordingly, then this opens the path from A to B, making them dependent given evidence for C. This is also intuitive. If we know the effect *and* we know the state of one of the possible causes, then we can reason about the state of the other cause. In network terms, if we know that "Battery" state is "low" and we have evidence that "Engine Cranks but w/n Start" is "Yes," then we can reason that "Fuel System" is less likely to be the cause. This is known as "explaining away" and is a powerful tool for articulating expert knowledge. The collider structure translates expert's statements such as "there are multiple causes for the engine to crank but not start" into a model that reasons like an expert; that is, if A is true and we know C, then it is unnecessary to test B (Figure 1.11).

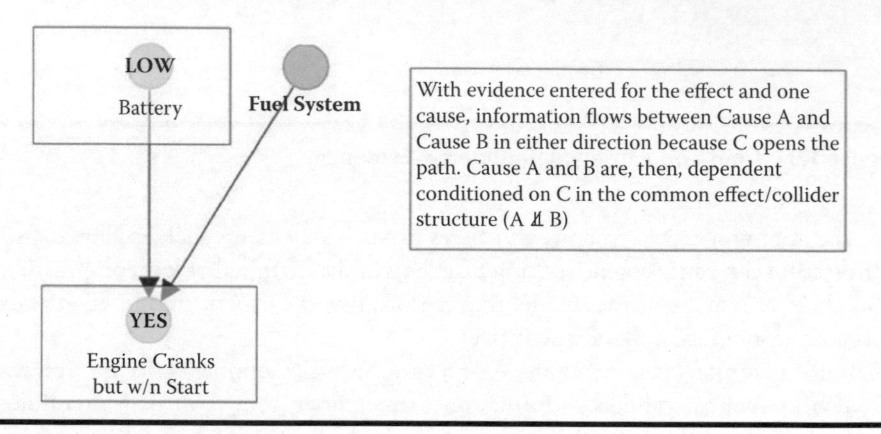

Figure 1.11 Common effect (collider)—Evidence.

The key takeaway from this section is that ANY piece of expert knowledge (variables and relationships between variables) can be encoded using these three structures, or combinations thereof, by the knowledge engineering team. We now have all the tools we need to develop the qualitative, or structural, component of BNs: the DAG. The BN is incomplete, however, without a means to specify the relationships.

The quantitative component of the CBN is constructed using conditional probability tables (CPT) and conditional probability statements for each parent-child relationship. These statements specify probabilities of each state of a child node given the state of its parents (Figure 1.12).

Node Editor

Node Selection: Effect Rename

Values | State Names | Reference State | Filtered State | Comment
States | Probability Distribution | Properties | Classes

Probabilistic | Deterministic | Tree | Equation

Cause 1	Cause 2	False	True
False	False	95.000	5.000
	True	60.000	40.000
True	False	40.000	60.000
	True	20.000	80.000

Complete Normalize Randomize

Accept Cancel

Figure 1.12 Conditional Probability Table (CPT): Quantitative component of a Bayesian network.

We will discuss, in detail, some techniques for working with CPTs in the Elicit section of DSEV. For now, it is easiest to consider this analogy when thinking about the quantitative portion: experts use the DAG to model what they think the world of their domain looks like; experts use the CPT to model how they move through that world. The qualitative and quantitative components combine to make a BN. Note that all of the structures described earlier are in use in the small network shown below (Figure 1.13).

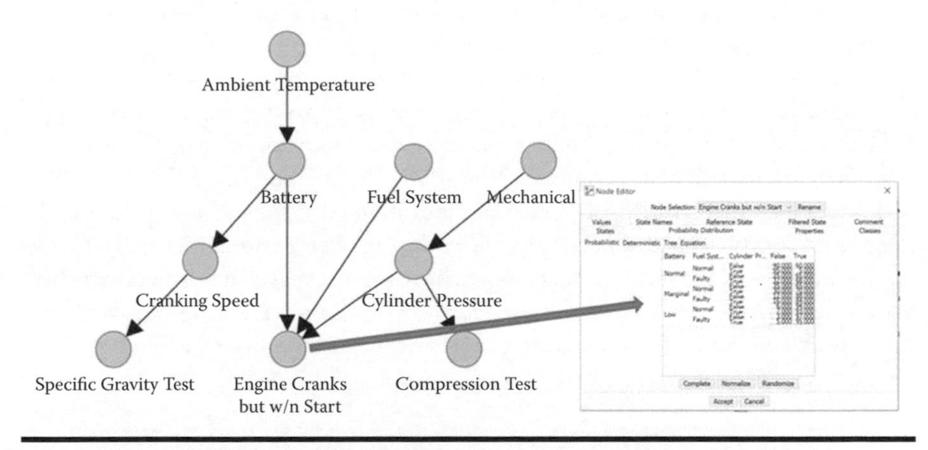

Figure 1.13 DAG and CPT: Qualitative and quantitative components of a Bayesian network.

The experts need to understand the basics and building blocks of BNs, need to know what their knowledge looks like in a BN, and need to see how the BN changes in response to different evidence. This is usually accomplished by building a small sample BN—6–8 nodes, as above—for a familiar domain and demonstrating how it changes, what structures are used to model different types of relationships, and what biases are in play. The BN in Figure 1.13 draws from our use case. While it is very basic, it demonstrates how easily relationships can be articulated and how it can frame further development. The translation of expert knowledge is such that the completed BN looks like, behaves like, and learns like the experts that contributed to its construction. With our experts now knowledgeable of the process and the tools, we move to a few other pre-DSEV tasks: framing the problem, identifying a target node, and generating the list of nodes (variables) that will provide the input for our DSEV process.

Pre-DSEV

This initialization stage starts with creating one to two very descriptive and specific sentences that fit under the objective and the use cases and that will guide the facilitation and model building. For our use case, we could use, "The generator type is a 60 kilowatt (kW) B model. The generator's diesel engine will crank (turn over) but not start in high ambient temperature." Using that as a guide, the symptom "diesel engine cranks but will not start" serves well as a target node. A note on target nodes: the goal is to find the right magnification. Too small, or narrow, and a target (i.e., Generator fuel pump inoperative) may miss many possible causes of the engine start problem; too large, or too wide, and a target (i.e., the Generator won't run) will cause the network to be too large and more difficult to elicit. A good target finds the middle ground with sufficient specificity to identify all problems but with a manageable network size. The experts are then asked to develop a list of nodes that influence, or are influenced by, the target node. A number of good, CTA-based techniques are useful for this process. Nominal group technique (NGT) is widely used because its silent voting process eliminates many of the biases associated with expert elicitations—most importantly, those that involve personalities influencing the development of the node list (e.g., more charismatic experts may influence other experts). In our use case, we would expect to see nodes representing causes of the engine problem and nodes representing effects of the engine problem (Figure 1.14). We are now ready to start the DSEV process proper.

Node List: Ambient Temperature, Engine Cranking Speed, Compression, Battery,

Fuel System, Specific Gravity, Mechanical, Cylinder Pressure

Figure 1.14 Node list.

Define

The node list is voted on, prioritized, and then discussed at length, removing, by consensus, duplicate or similar nodes. The imperative of the Define phase is to ensure the state space in each node meets all criteria. The state space is a simple description of what values the variable can take on in nature. The state space must be exhaustive (all possible states enumerated) and mutually exclusive (the variable can only be in one state). Most importantly, the state space must be intuitive to the KE team and to the experts. Our CBN tools provide some help in that regard by enabling us to add contextual information on values to establish thresholds, assign upper and lower bounds, and make the states more specific. These can be as simple as binary state space (true–false; yes–no), or multi-state where the value field provides the necessary threshold levels (Figure 1.15).

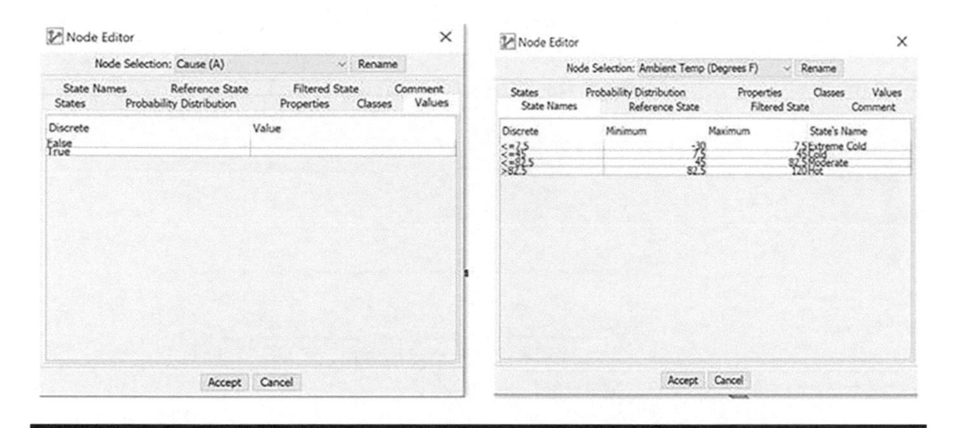

Figure 1.15 State space.

The "clairvoyance test" is unique to DSEV and provides the mechanism by which we can test our work on developing state spaces. The premise of this test is that if a person with perfect information provided a number or name value of a variable, would this value make sense given only the variable name and state description? In our "Ambient Temperature" example earlier, a clairvoyant value of "28" would be easily translated into the "Cold/7.5–45 degree" state. The primary goal of the KE team and experts is to remove vagueness and ambiguity from the state space. With this thoroughly vetted list of variables in hand, we are ready to start the Structure phase. In this next section, we will discuss how to categorize node types and how to integrate the nodes on the list in a network.

Structure

The first step in the developing the DSEV Structure phase is to work with a small "chunk" of nodes and connect them. This requires knowledge of the three basic BN node structures (discussed previously) and of the different types/categories of nodes in a BN. BNs consist of five types of nodes: hypothesis nodes, context nodes, indicator nodes, report nodes, and intermediate nodes. The state space of all types of nodes must be, as discussed in the Define phase, mutually exclusive (the node can only be in one state) and exhaustive (the state space must cover all that are possible). We will briefly describe the attributes of each.

Hypothesis nodes describe possible root causes for the symptom. Hypothesis nodes precede indicator nodes and report nodes in temporal order (they are parents of indicator and report nodes). This means directed arcs go from the hypothesis node to the indicator or report node. Hypothesis nodes succeed context nodes in temporal order (they are children of context nodes). This means directed arcs come into the hypothesis node from the context node. "Battery," "Fuel System," and "Mechanical" are examples of hypotheses nodes for our problem statement (Figure 1.16).

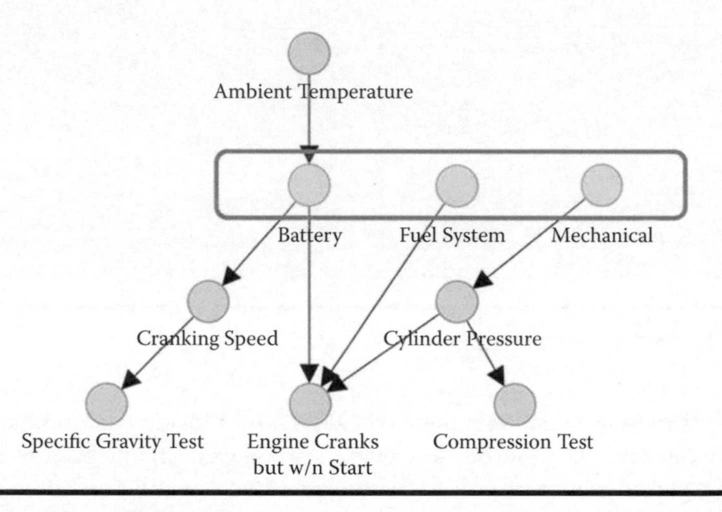

Figure 1.16 Hypothesis nodes.

Context, or pre-dispositional, nodes precede the hypothesis node in temporal order (they are parents of hypothesis nodes). This means directed arcs go from context nodes to hypotheses nodes. "Ambient Temperature" is an example of context for the "Battery" hypotheses, as it could predispose the failure condition (e.g., cold temperatures are correlated with low battery cranking performance) (Figure 1.17).

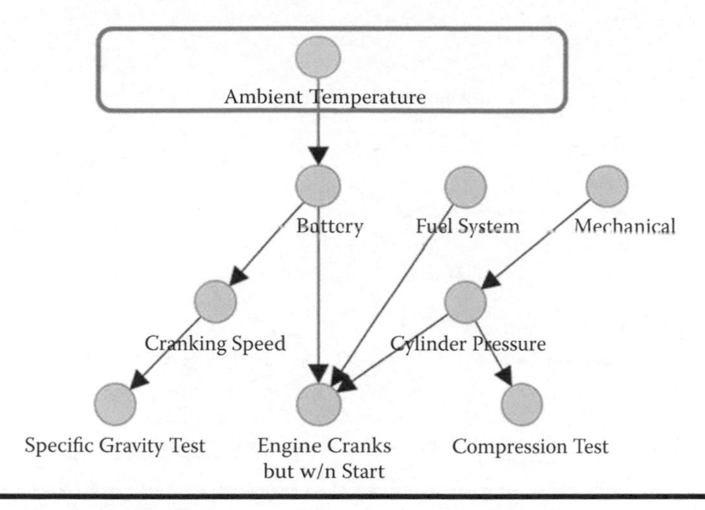

Figure 1.17 Context nodes.

Indicator nodes are children of hypothesis nodes. Directed arcs come into indicator nodes from hypothesis nodes. Indicator nodes are, in general, symptoms of the hypothesis node, but they may not always be easily observable. "Cranking Speed" and "Cylinder Pressure" are examples of indicators of the hypothesis, as they are symptoms but not readily observable (Figure 1.18).

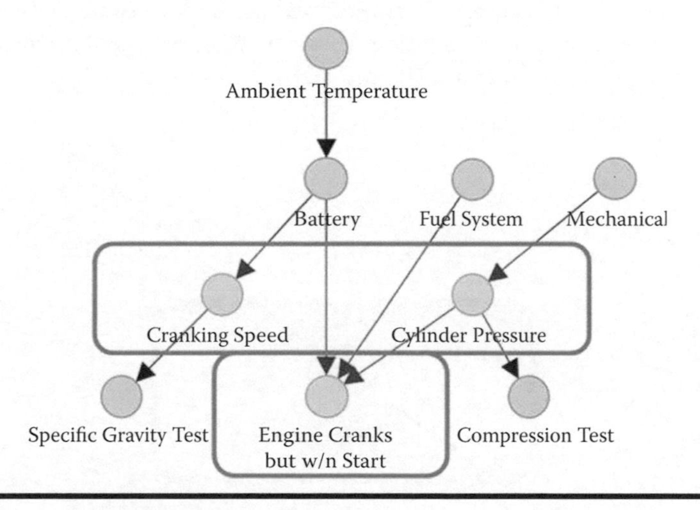

Figure 1.18 Indicator nodes.

Report nodes are children of indicator nodes. Directed arcs come into report nodes from indicator nodes. Report nodes, in general, are observables of the indicator node. "Specific Gravity Test" and "Compression Test" are examples of report nodes for the hypotheses as they are both observables for the indicator node (Figure 1.19).

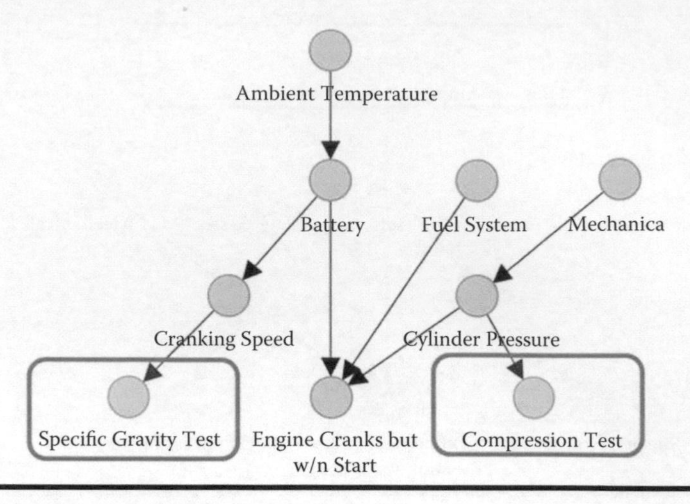

Figure 1.19 Report nodes.

Intermediate nodes—a part of the indirect connection structure described previously—can be used with all types of nodes as the name suggests; for example, context nodes can influence the hypotheses through an intermediate node, hypotheses nodes can have intermediate effects between them and indicator nodes, and so on. All of the same rules apply to intermediate nodes. "Cylinder Pressure" would be an example of an intermediate node between "Mechanical" and the hypothesis, as it describes one of the mechanisms whereby mechanical issues affect engine start functions (Figure 1.20).

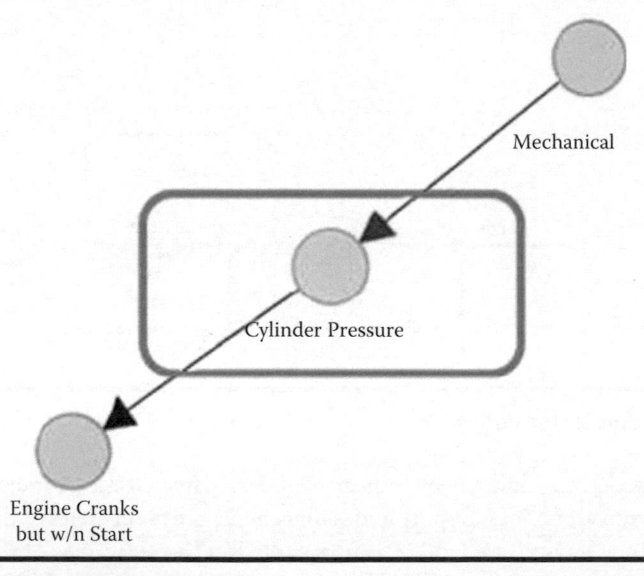

Figure 1.20 Intermediate nodes.

Another example from our "diesel engine cranks but will not start" problem statement ties all these node types together: Cold ambient temperatures (predisposition) slow down the flow of electrons through a battery's electrolyte (hypothesis) causing slow engine cranking speed (indicator, not directly observable/quantifiable) that can be detected using a specific gravity test (report) (Conrady and Jouffe, 2015) (Figure 1.21).

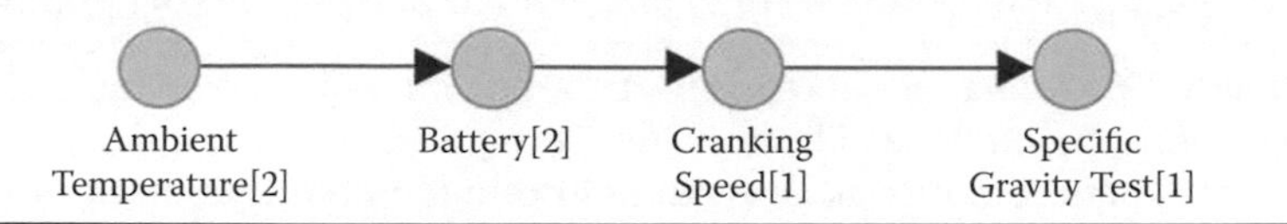

Figure 1.21 All node types in action.

The key knowledge engineering task of the Structure phase involves deciding what types of nodes we have in the list from the Define phase, where those nodes fit in the model, and what structures to use to represent the knowledge. With all this in hand, we can move to the description of the Elicit phase, where the structural relationships described previously can be specified.

Elicit

This phase involves drawing out from the experts the probabilities for the structural model you have built in the previous phase. This estimation process is burdened with a number of biases that the team must be (a) aware of, and (b) have a mitigation plan for. Before outlining the soft and hard skill requirements for the Elicit phase, Table 1.2 is provided as a quick reference on common biases and mitigation techniques.

Table 1.2

Anchoring: Initial impressions, whether quantitative or qualitative in nature, can cause bias as the human brain tends to give more weight to the first information received.
Mitigate by making the team aware, by avoiding using specific numbers or specific contexts, and by providing a multi-perspective background.
Availability: The human mind tends to respond with information supported by what information is readily available. This means that particularly "sticky" recall (traumatic, emotional, or vivid) of information tends to bias estimates.
Mitigate by making the team aware and attempting to take emotion out of the estimation process by reviewing pertinent statistics.
Confirmation: The human mind tends to seek out confirming evidence. Given a preferred outcome, an individual will tend to seek out evidence to confirm that outcome, and interpret evidence to confirm that outcome, instead of giving equal weight to evidence that conflicts with the desired outcome.
Mitigate by making the team aware and actively seeking out disconfirming evidence. Try to draw out scenarios that would result in a different conclusion.

With this knowledge in hand, we move our focus to the key DSEV techniques for eliciting probabilities for each node. One of the key challenges in this process is in establishing the base rate, or prior probability, for each node. The base rate should be an estimate of the state in nature with no other forces acting upon it. For example, when eliciting the prior probability of a "cranks but will not run" fault, consider some relevant parametric data, such as your own automobile. Does the engine fail to start 1 out 100 starts? 1 out of 1000? 1 out of 10,000? Consider these ratios knowing the tendency is to set the base rate too high. Consider both data *and* context. For example, the base rate for smokers vs. non-smokers in a chest clinic is significantly different that the base rate for smokers in a random selection of people on the street. The prior probability should be entered into the nodes conditional probability table (Hepler et al., 2015) (Figure 1.22).

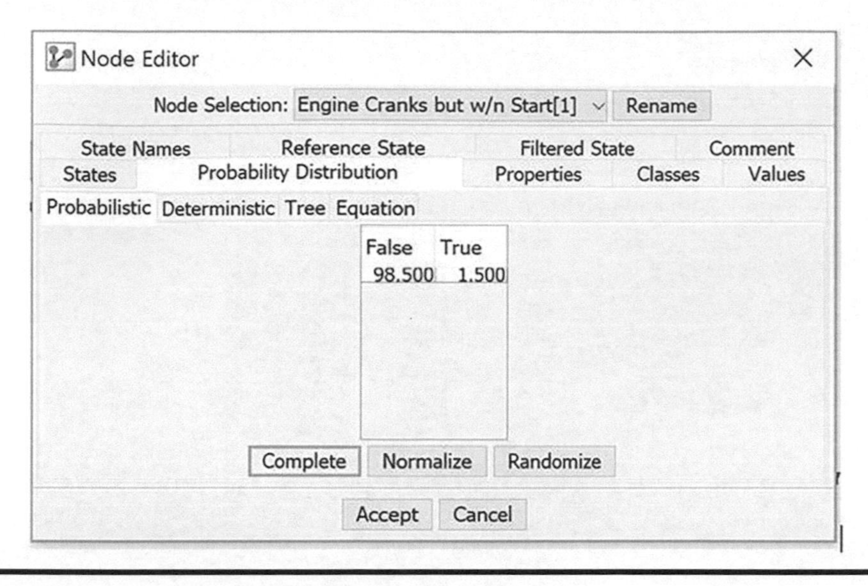

Figure 1.22 Conditional Probability Table (CPT).

For root nodes (nodes with no parents) these probabilities can be entered directly into the CPT. For any node with parents, the node must be temporarily disconnected (arcs coming into the node disconnected) to input prior probabilities. This should be done for all nodes before reconnecting all the arcs. The resulting reconnected network will show the marginal probabilities for each node, or the probabilities with no evidence entered for any node, but the nodes will indicate that the underlying CPT still needs to be filled in. As detailed earlier, the CPT is a series of statements that show the probability distribution for each state of a child node given each state of its parents (Conrady and Jouffe, 2015) (Figure 1.23).

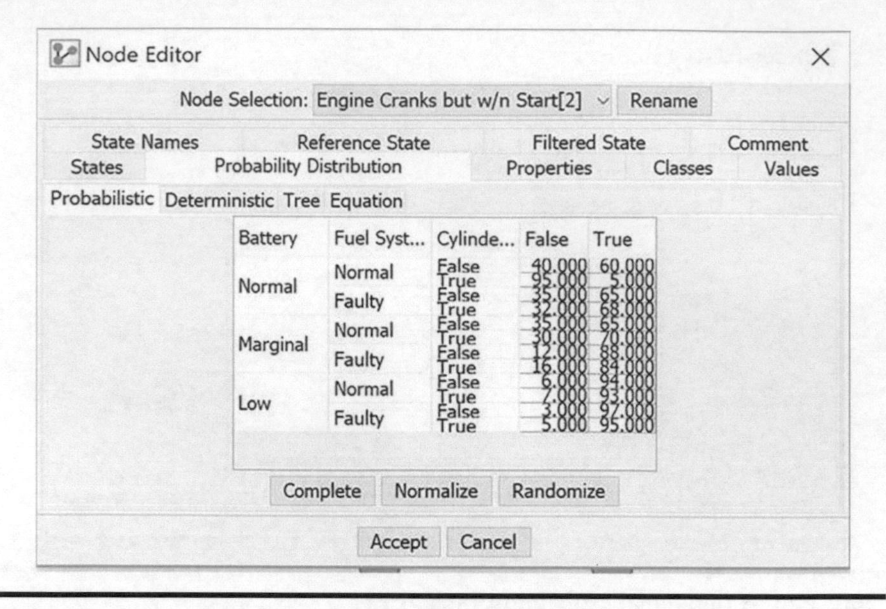

Battery	Fuel Syst...	Cylinde...	False	True
Normal	Normal	False	40.000	60.000
		True	95.000	5.000
	Faulty	False	35.000	65.000
		True	32.000	68.000
Marginal	Normal	False	35.000	65.000
		True	30.000	70.000
	Faulty	False	12.000	88.000
		True	16.000	84.000
Low	Normal	False	6.000	94.000
		True	7.000	93.000
	Faulty	False	3.000	97.000
		True	5.000	95.000

Figure 1.23 Conditional probability statements.

The knowledge engineering team and the experts work through each of the statements. This is the heart of the process as these estimations are what makes the expert knowledge show through in the BN and what makes the network think and act like an expert. There are many direct and indirect methods for drawing out these estimations. Experts like to work with words. Direct methods use questions designed to elicit a number response; indirect methods use questions designed to elicit relationships that can be converted to numbers (Table 1.3).

In some cases the team may find it easier to use a formula, such as a normal distribution, to translate expert statements and fill in the table. The key techniques here are to avoid bias and to use tools to convert language into probabilities. The elicited probabilities can be "played back" to the experts in real time to ensure they agree with how the network responds to evidence. This is the essence of the final phase (Hepler et al., 2015).

Table 1.3 Direct and Indirect Methods for Estimating Probabilities

Method	Query
Direct Method	What is the probability of an event (there will be 2+ inches of rainfall) taking place?
Fractiles (percentiles)	What is the probability that rainfall will be less than or equal to 2"?
Quantiles (inverse of fractile)	What amount of rainfall is associated with .5 probability (equally likely to be above or below that amount of rainfall)?
Odds	What are the odds (1 chance in ?) that there will be 2" or less of rainfall?
Analytical Hierarchy Process	How many more times likely is it for there to be less than 2" of rain than there to be more than 2" of rain?
Confidence Intervals	How many times out of 1000 samples would less than 2" of rain fall?
Words with numbers attached	Which of these words best describes your uncertainty about whether less than 2" of rain will fall: certain (100%), probable (75%), fifty-fifty (50%), unlikely (25%), or never (0%)?
Standard gambles	Is the probability of less than 2" of rainfall greater or smaller than drawing a club from a deck of playing cards?
Probability wheel	Is the probability of getting less than or equal to 2" of rainfall greater or smaller than the pointer wheel landing on a designated color?
Rank events then convert to probabilities	Rank order events from most likely to least likely then use a rank order centroid formula to calculate probabilities
Conjoint analysis	Establish rank order relationships (more likely, equal, less likely) between combinations of events and use these relationships to infer a set of numbers that meet constraints

Verify

The sub-headings under this phase—consistency, coherence, what-if analysis, and sensitivity analysis—all have to do with ensuring that the model you are working on is operating as intended. Working with your expert team, use these techniques to exercise the model in different ways to uncover any possible discrepancies, or to just fine-tune the model. Consistency tests involve going into the CPT of each child node, rearranging the order of the parents in all possible combinations, and observing patterns after each reordering. Note below that rearranging the order of parents makes it more obvious to the experts that the relationship between "Cylinder Pressure" and "Engine Cranks but w/n Start" is independent of the other parents. This will show, at a wide angle, how each parent node affects the child node distribution, and this provides one way for the experts to review their work and acknowledge the veracity of the parent-child relationships (Figure 1.24).

Similarly, coherence tests just provide the experts and the team with a different perspective on the model. By examining the marginal distributions (no evidence entered), we can analyze the relationships between variables and see if there is good fit between the expert's knowledge and the model. In most cases, this is a way to

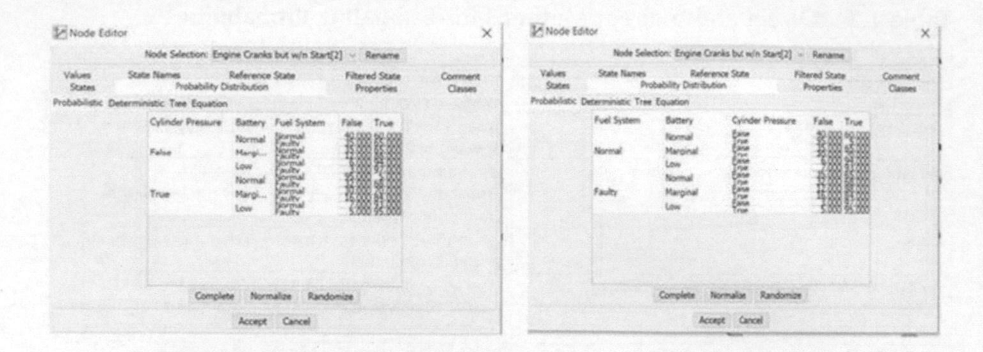

Consistency: Does the CPT make sense when varying context or causal variables?

Figure 1.24 Consistency.

Coherence: Do the marginal distributions make sense?

Figure 1.25 Coherence.

verify that the prior probability distributions and the conditional distributions in the CPT are representative of the expert's thought processes (Figure 1.25).

The "what-if" tests, as the name implies, involves "playing" with the model by entering different sets of evidence and seeing how the model acts under different scenarios. Always prepare experts for this test by asking what they *expect* to see. The perspective on elicited results and computed results (2) provides yet another way for experts to think about their view of the domain (Figure 1.26).

The key technical takeaway for this phase is mutual information. In building the model, we are interested in how much entropy is reduced in a variable by instantiating evidence in other nodes in the model. Mutual information sensitivity nests well with the "first best question" philosophy that guides the practical tool for soldiers and diagnostics in general. We want to eliminate as many possibilities as

What-if: Does the model work as expected when entering evidence?

Figure 1.26 What if?

Mutual information: How much does one random variable tell us about another?

Figure 1.27 Mutual information.

possible in the shortest amount of time, and we want to be cognizant of opportunity cost with respect to alternate diagnostic paths. BN software packages provide easy-to-use tools for this type analysis (Hepler et al., 2015) (Figure 1.27).

We will now fit a DSEV elicitation around our generator use case. Diesel engine-driven generators are the life-blood of any combat operation. Generators provide clean, reliable, and quiet power to critical weapon systems and life-support systems. The generator itself is a complex piece of machinery, with electrical, electro-mechanical, and electronic components all integrated for diesel engine operation, electrical power production, and control and safety circuits for both. The Army has limited data on maintenance of these generators in the field. What exists is mainly transactional supply records that are only weakly associated with failure modes. What's more, this weak connection results in suboptimal supply policies and provides no actionable data for soldiers at the point of need. What the Army does have is a very rich source of experiential knowledge from the field support personnel that are tasked to support these systems in the field. In some cases, this knowledge is made explicit in the form of quick reference guides and information papers; in some cases, this knowledge is tacit. In all cases, this knowledge is most important to tap into. Using the combination of DSEV and BNs, expert knowledge will be transformed into a data-generating process that starts with accurate diagnosis and flows through the enterprise as accurate data with which to make supply and repair decisions.

Critical Decision Method (CDM)

Our group of experts includes personnel with specific experience with supporting generators in deployed areas, along with personnel with general experience in deployed areas and personnel with engineering knowledge on the generator sets. This section will show how CTA methods are woven into the DSEV to produce BNs that "think" like this group of experts. Getting to this point, however, requires a means for putting these experts "in the moment" and then taking advantage of that opportunity with a structured process for creating an explicit, time-ordered account of the experts' reasoning about particularly critical problems. One of the most effective methods for doing this is the critical decision method. CDM involves a number of elicitation "sweeps," each with a different purpose but overlapping in key aspects (Hoffman et al., 1998) (Figure 1.28).

The art of CDM is in selecting the right event. In our case, we want to put our power generator experts back into their deployed experiences: ones where they were the decision makers, where their decisions mattered, and where the situations were unusual or significant enough to create vivid memories for the experts. In the power

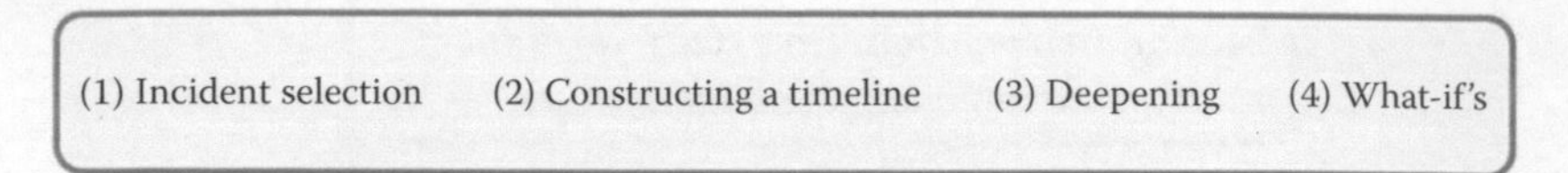

Figure 1.28 Critical decision method "sweeps."

generation world, these cases will usually involve calls for assistance where critical systems are down due to a lack of power and the usual supply and maintenance resources are not available. For example, in Iraq in 2003, water purification assets were placed strategically along convoy and supply lines to ensure coalition troops and civilians had a dependable source of potable water (Figure 1.29).

These assets were often located on river side farms owned and run by local Iraqi families. The only power sources sufficient to run the water purification systems were the diesel generators mounted on the system trailers. Failure of these generators and the loss of purified water production put soldiers and civilians at immediate risk. Our specific case involves a generator deployed to a farm along on the Tigris River in Iraq (Figure 1.30).

Figure 1.29 Tactical water purification equipment.

Figure 1.30 Tactical water production in Iraq.

During operation, an electrical short had caused catastrophic failure of the generators Input/Output (IO) module—the "brain" of the generator control system. This caused the generator to shut down during operation with no specific shutdown indicators. A field support person, one of our experts, had only a matter of hours with which to travel to the affected site via convoy, diagnose the problem, and effect short- and long-term repair plans. This scenario meets all the criteria for incident selection.

The second sweep in CDM involves an iterative process of developing a timeline of events that covered the incident. In our case, we want to know, from the time the expert was notified of the problem, what were his or her reactions, preparations, and thought processes, and how did those translate into an ordered approach to addressing the problem—remembering that the combat environment introduces uncertainty into the basic process of getting to the site. From this process, we gather the following:

1. The expert convoyed to the site and was delayed several times along the way due to enemy activity.
2. Without any communication with the site during the convoy, the expert started running possible cause and effect relationships and started developing a set of questions.
3. Upon arrival at the site, the expert asked key questions of the crew to create a sub-timeline of the failure itself, e.g., did it occur during full load operation, no-load operation, starting, etc.

From a DSEV perspective, the knowledge engineering team uses this information to confirm a target node and to compile a list of these potential "first-best" questions that will become context and indicator variables. As the timeline gets richer in detail, we gather that the expert had a good idea of what he or she wanted to look at first when gaining access to the generator, and this was an "audible click" test, not available in any technical manual or doctrine, that the expert himself had developed from his own experience and from discussions with engineers. The test was designed as a quick way to determine if power was available at the fuel injection pump actuator. A positive test result (audible click) would verify all control circuits were functioning and would place the problem in the mechanical or fuel supply subsystems. A negative test result would mean that a failure in the control circuit was nearly certain. The test effectively isolates the problem to mechanical or electrical and reduces a significant amount of entropy with very little time or effort required (Figure 1.31).

Per our DSEV process guidelines and based on our expert team make-up, the majority of these experts will have similar experiences troubleshooting critical failures in deployed areas. The discussion of the audible click test will stimulate discussion on if, and how, the test should be a part of the troubleshooting network. It will also stimulate discussion on what the other experts do in the same situation, possibly revealing other such tests. These are exactly the types of discussions DSEV can

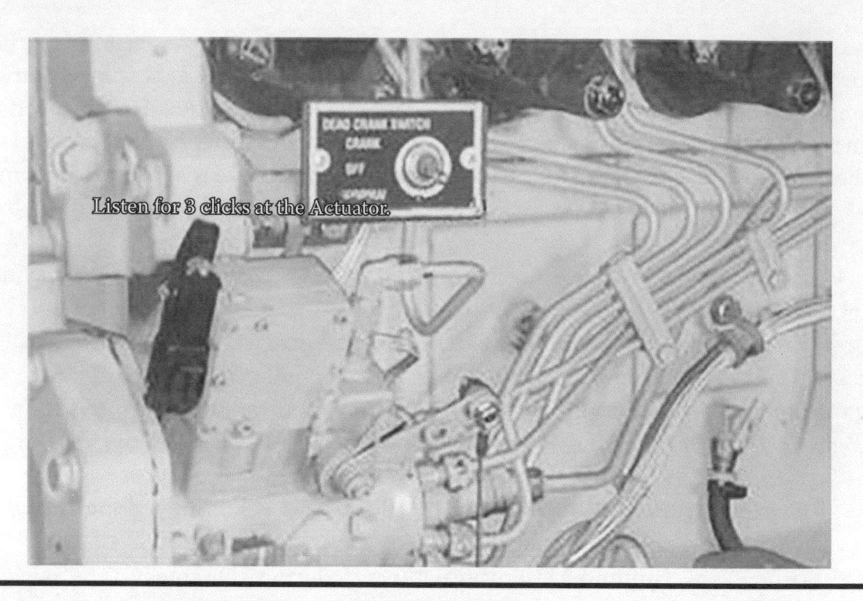

Figure 1.31 Diesel generator audible click test.

facilitate. With all the stress of the combat environment and situation, this particular expert chose this tack (this test approach) in the most extreme circumstance because he or she trusted it would reduce the most amount of uncertainty. This is significant as the fact that this particular expert put this particular test in play at this particular time is an extremely valuable insight into his (her) expertise. But we want to know more about the thought process.

The "deepening" sweep is intended to fill in the blanks from the expert(s) statement "I just knew," or "I had a gut feeling," about a specific action. The key to executing this sweep in DSEV is to ask questions that help jog the expert's memory about the cues and mental models that they use. The elicitation question, "What were you seeing when this occurred?" is an effective way to start this process of drawing out and understanding the key cues from each event in the timeline. Other experts can help in the memory recall by each providing options from technical jargon, for example, was this light illuminated, did you hear this relay energize, what was the switch configuration, and so on. Quickly, we piece together the five-sense indicators available to the expert and how the expert processed the information. Returning to our case, the expert was given information upon arrival about how long he had to diagnose the problem. He was advised that there was another failed generator at a different location and that the convoy would be proceeding to that location in less than one hour. Missing that convoy would mean the second site would be without power for a minimum of 24 hours (before the next convoy). The expert immediately decided to attempt a diagnosis and field-expedient repair at the original site to restore power and then make the convoy to the next site. This gave him less than 30 minutes to complete the diagnosis and repair. Given this schedule, the expert looked at three visual indicators prior to making the decision to proceed with the audible test: engine crank (yes–no), I/O "heartbeat" (a software–hardware communication indicator), and an engine speed indicator signal (Figure 1.32).

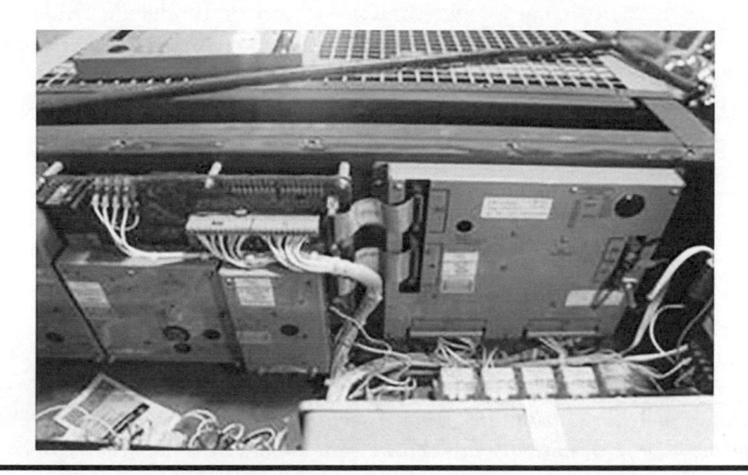

Figure 1.32 Visual indicators support troubleshooting decisions.

The knowledge engineering team uses this information to narrow the node list for a specific troubleshooting path and to sort those visual indicators as formal nodes in the network. The deepening sweep continues in this manner through the timeline, iteratively, to ensure no key pieces of information are missed or out of sequence. In these first three CDM steps, the team has developed the core of a BN infused with expert knowledge. It remains, then, to discuss if changing the states of some variables could have impacted the decisions.

The "what-if," or "hypotheticals," sweep is intended to make the experts change perspective and think of ways the model could be improved. Some of the key questions to ask here are, "What is the difference between the way you approached the problem and the way a novice would have approached it?" or, "What information would have helped you make a quicker diagnosis?" In our generator case, hewing to the time-critical aspects of the scenario, the "what-if" sweep results in extraction of several other time-saving tests that could be used individually, or in combination, to isolate the problem more quickly. CDM sweeps can and should be run multiple times for the same problem set, or until the expert team is satisfied that the model is correct. Many other CTA-based techniques exist, but for them to be effective in a knowledge elicitation process, they all must be grounded in significant events where experts were fully involved.

We have sketched a process for extracting and codifying expert knowledge into a practical and useful model for soldiers that can support diagnostics, decision making, analysis of alternatives, and myriad other analytical processes.

The tools recommended herein are feature-rich—we have only scratched the surface of capabilities, techniques, tools, and details that can be brought to bear on any problem set. DSEV itself encompasses a library of best practices in knowledge elicitation, too many to detail here. Bayesian Belief Network's state-of-the-art software is such that virtually any problem set can be modeled, validated, and analyzed using the most advanced operations research techniques. We recommend this reading as an introductory guide based on multiple, quantifiable successes using DSEV and BN development software in the field. Readers are encouraged to delve further into DSEV and the use of BNs to shape a customized problem-solving approach.

References

Aebischer DA, Vatterott Jr. J, Grimes M, Vatterott A, Jordan R, Reinoso C, Baker BA et al. (2017) Bayesian networks for combat equipment diagnostics. *Interfaces*, 47, 85–105.

Bilmes LJ (2013) The financial legacy of Iraq and Afghanistan: How wartime spending decisions will constrain future national security budgets faculty research. Accessed September 20, 2017, http://watson.brown.edu/costsofwar/files/cow/imce/papers/2013/The%20Financial%20Legacy%20of%20Iraq%20and%20Afghanistan.pdf.

Conrady S, Jouffe L (2015) *Bayesian Networks and BayesiaLab: A Practical Introduction for Researchers* (Bayesia USA, Franklin, TN).

Hackman JR, Vidmar N (1970) Effects of size and task type on group performance and member reactions. *Sociometry*, 33, 37–54.

Hepler AB, Tatman JA, Smith GR, Buede DM, Mahoney SM, Tatman SM, Marvin FF (2015) Bayesian network elicitation facilitator's guide. Report, Innovative Decisions, Vienna, VA.

Hoffman RR, Crandall B, Shadbolt N (1998) Use of the critical decision method to elicit expert knowledge: A case study in the methodology of critical task analysis. *Human Factors*, 40, 254–276.

Kiesling EC (2001) On war without the fog. *Military Review*, 81, 85–87.

Korb KB, Nicholson AE (2011) *Bayesian Artificial Intelligence* (Taylor & Francis, Boca Raton, FL).

Pearl J (1982) Reverend Bayes on inference engines: A distributed hierarchical approach. Accessed September 19, 2016, https://www.aaai.org/Papers/AAAI/1982/AAAI82-032.pdf.

Pearl J (1988) *Probabilistic Reasoning in Intelligent Systems: Networks of Plausible Inference* (Morgan Kauffmann, San Francisco, CA).

Pearl J (2015) *An Introduction to Causal Inference* (CreateSpace Independent Publishing Platform, Seattle, WA).

Chapter 2

Network Modeling and Analysis of Data and Relationships: Developing Cyber and Complexity Science

Chris Arney, Natalie Vanatta, and Matthew Sobiesk

Contents

Introduction

Networks are not only ubiquitous, but also lie at the core of the economic, political, military, and social fabric of modern society. As stated in a National Research Council (2005) report, "society depends on a diversity of complex networks for its very existence." Today's scientists and analysts perform network modeling and data analytics in many applications. Network science (NS) as a component and partner of data analytics (DA) is critical to understanding many military-relevant issues, such as (Brandes et al., 2013):

- Communication flow
- Command and control
- Managing unit operations
- Implementing information assurance
- Modeling terror cells and their processes
- Gathering and processing intelligence
- Maintaining security in physical and informational systems
- Enabling engagement within the Army's Global Landpower Network
- Decision making

In health and biology, network applications include mapping genetic and protein networks and their roles in disease, representing the brain morphology by networks, forecasting and diagnosing disease contagion, and analyzing various levels and regions of ecosystems. Network-based applications involving physical networks include managing communication and computer systems, operating logistics and transportation systems, and designing infrastructures in various buildings, structures, and systems (e.g., water, waste, and heat). The network models associated with social processes involve conducting collaborative decisions and group learning, modeling collective/team behaviors that involve coordinated activities, modeling the meanings of textual and spoken language to understand influence, and tracking the emergence of societal impact and achievement.

Network models can merge the social, informational, communication, and physical layers of a system or organization into an interconnected, unified system for a broad, integrated analysis (Arney, 2016; Arney and Coronges, 2015). The network layers collectively produce an all-encompassing, non-reductive model that permits multiple scalings, processing of tremendous volumes of data, suitable system complexity, and appropriate diversity and specializations within the system or organization. Network modeling enables the dynamics in the structures and processes of the phenomenon being modeled to build usable knowledge and results for viable decision making. One of the recent results in NS shows the impossibility of perfect control over organizational and entity behavior in social networks (West, 2015). A subtle hand of delicate management through shared vision and autonomy is often more powerful than rigid micro-control through rules, regulations, and detailed instructions. This result is significant for military leader networks. As Roehner (2007) reflected about these data-driven contexts: "the real challenge is to do real physics and real sociology in the framework of network theory."

Data analytics (DA) encompasses even more techniques and methods than NS. The National Academies (2017, p. 64) describes and categorizes DA as:

- Descriptive and exploratory: "Using data to summarize and visualize the current state."
- Predictive: "Using data to determine patterns and predict future outcomes and trends."
- Prescriptive: "Using data to determine a set of decisions … that give rise to the best possible results."

The techniques used to perform these DA components are data extraction, natural language processing, data mining, machine learning, statistical analysis, stochastic modeling, regression, optimization, simulation, clustering, and classification. The combination of both unstructured text and highly structured quantitative data sometimes is referred to hard-soft data fusion.

We will report on network modeling and data analytics as they relate to two basic-science issues with associated applications within the military—cyber science and complexity science. In particular, we will outline the type of DA and NS work being performed and provide examples in both areas. Cyber science often relies on descriptive and predictive data analytics to find anomalies in active network processing and prescriptive data analytics to determine actions needed to protect or fix attacked networks. The U.S. Department of Defense (DoD) Cyber Strategy (2015) guides much of the efforts we have assembled in our modeling and problem solving of cyber and information science problems.

Cyber Science

One area where we use NS and DA is in cyber problem solving. We use network modeling to build a framework for cyber problems using game theory concepts linked to mathematical topology (information networks) and cultural modeling (social networks).

In 2006, the Joint Chiefs of Staff decreed the Fifth Operational Warfighting Domain for the United States military: the cyber domain. At times, a seemingly made-up word, cyber is still not clearly defined and understood a decade later. Academia, industry, popular media, and government can all debate the correct usage of the noun, but for purposes of this discussion, we will use definitions from the military community. *Joint Publication 3-12 (R) Cyberspace Operations* of the U.S. DoD (2013) defines cyberspace as "the global domain within the information environment consisting of the interdependent network of information technology infrastructures and resident data, including the Internet, telecommunications networks, computer systems, and embedded processors and controllers." The Joint Publication further comments that cyberspace has multiple layers: "physical network, logical network, and cyber-persona" layers. The physical network layer consists of the geographic components that produce the medium where the data travel. The logical network layer is an abstraction of the information space (e.g., the URL for a website, no matter where it resides in the physical network). The cyber-persona layer represents an even higher level of abstraction of the logical network (e.g., the website itself and the people that build and operate the network). As the military continues to explore the cyber domain (both with an offensive and defensive mindset), the principles of network science and the analysis of data play a large role in their understanding.

In general, cyber operations are not constrained by geographic terrain but are bounded by the network connections and the cultural space of the operational mission and forces. Often the topology of the network dramatically affects the operational plan and situational awareness. Primary concerns in cyber situational awareness are defining information needs and the types of activities involved in the operation. The topology of the network models for information operations is much like a terrain map or operational overlay for physical operations. Understanding the cultural domain involves analysis of human and technical factors, such as computer and network platforms, cultural bias, ability to collect relevant and timely data, and legal issues. On the analytic side, the technical work involves analyzing the data to show the connections of structures and processes, and identifying possible patterns in the dataset. The methodological tools in network modeling including machine learning, qualitative and quantitative analysis, and statistical evaluation (Carter et al., 2014; Mayhew et al., 2015).

Today's military cyber forces have three primary missions: (1) defend DoD networks, systems, and information, (2) defend the U.S. homeland and U.S. national interests against cyberattacks of significant consequence, and (3) provide cyber support to military operational and contingency plans. Under these missions, cyber forces must be able to understand the current network infrastructure that is used both for command and control capabilities but also for watching the latest cat videos on YouTube. See Figure 2.1 for a framework for such a network structure.

One interesting challenge in understanding the infrastructure is the development of a PACE (primary, alternate, contingency, and emergency) plan for communications. These should be four separate means of organic communication capabilities that units can fall back through in case of transmission disruption. Traditionally, this might be secure FM radio, then high-frequency radio, then secure satellite phone, and finally cellular phone. With the many layers of abstraction in our evolving communications technology, it is even difficult for experts to understand the potential points of failure and interdependencies within transmission spectrums in these extremely complex environments. For example, there is a security operations center supporting multiple states that recently fell prey to this complexity. They built the center with extreme redundancies: dual ISPs providing bandwidth over two different fiber lines that entered the facility from completely different cardinal directions. They did everything to ensure separate communications paths to protect their operations, and yet it turned out that within 50 miles from their facility, both links went over a single bridge. NS modeling can be used to visualize and simulate the complex communications spectrum to ensure that

Figure 2.1 Interaction of cyber domain aspects.

military PACE plans are accurate and actionable. Examination of multimodal transportation systems uses a set of subnetworks (each representing a transportation method such as plane, train, car, boat, bicycle, and walking) and then builds a logical network to describe connections between the layers. These same techniques could be used to describe the layers of communication spectrums (such as cellular, high frequency [HF], ultrahigh frequency [UHF], wireless, satellite) and resulting physical infrastructure. Then, apply tools of percolation theory and cascading failures to determine viability of the PACE plan during operations.

Additional NS and DA skills can be focused on the data storage and processing problem. On January 10, 2017, Army Secretary Eric Fanning directed that 60% of the Army's 1200 data centers must be closed by the end of 2018 and 75% by 2025. This is an effort to consolidate services to improve security and workforce efficiencies. This translates into a network optimization problem spanning the globe. Where do you place certain services to cover the military installations across continents? A current data center could be providing 1–2 services or up to 40+ different services to individuals within a mile of the location or spanning continents. With different characteristics of services, how to design an optimized location of the remaining data centers that can still meet the operational needs for the military both when they are working in garrison and when they are deployed in the field is an interesting problem.

Given the complexity and ad hoc nature of most military network infrastructure, it becomes difficult to model and/or simulate the normal behavior of traffic (Paxton et al., 2014). Lack of robust modeling and DA capabilities hinders leaders' abilities to make strategic risk decisions, quantify the impact of degraded systems, and identify cyber key terrain. When troubleshooting communications networks, often the issue and problems that need to be addressed in the operational environment are hidden and only the symptoms are visible. The determination of the cause of the problem—a bug, an innocent human error, bad hardware, multi-layered firewall, a dynamic, fast-acting AI, or a malicious attack—is often undetectable without clever and innovative use of DA. DA can also provide insights into the optimal placement of sensors within networks to detect maliciousness without undermining efficient routing of traffic. NS techniques can also help reverse engineer the infection of networks by malware. Cyber analysis is needed for time-sensitive problem solutions and dynamic network performance that are not as common in other domains.

Complicating the cyber forces' defensive mission is the massive amount of data that is generated by networks, systems, and people. The search for the needle in the haystack (solution) is difficult due to the volume of noise. DA techniques coupled with behavioral understanding are required to turn data into information and then into actionable intelligence. A defender must look not only for the outside threat but also for the insider threat. Network traffic sensors (aimed at detecting abnormal patterns), physical security logs, system access logs, and many more data streams (both technical and social) are fed into "Big Data" platforms. Then, using machine learning techniques, a system can detect insider threat instances from a non-human perspective with surprising accuracy, drawing connections that were not previously considered relevant.

The intersection of large datasets and machine learning is causing an evolution of data analytics. A large information-based company processes 600 billion events a day (approximately 3 petabytes of data a day) through a single pipeline and then uses an assortment of machine learning techniques to sort events, whether they are processed in real time or batch processed at a later time. The system learns situational behavior to make autonomous decisions at any given moment in any given situation about what data should be immediately processed to provide the best real-time quality of service to customers. While the military insider threat problem described in the previous paragraph would result in significantly less than 600 billion events a day for a given network region, the ability to sort data at machine speed for various processes based on the developing situation would provide quicker indications of malicious actors on the network.

NS concepts also provide cyber defenders understanding of the network attack surface. Attack graphs as shown in Figure 2.2 represent the paths through a network that end in a state where the intruder successfully gains access to protected systems (Ingols et al., 2009). The size and complexity of modern networks make it difficult to manually determine/maintain an account of the exploitable surfaces. Researchers are exploring new techniques and algorithms to overcome the complexity of the state explosion problem, which occurs when the number of vulnerabilities in a network grows large. Coupling attack graphs with machine learning capabilities allows for models to also hypothesize on missed vulnerable paths and/or missed alerts that intrusion detection systems don't see.

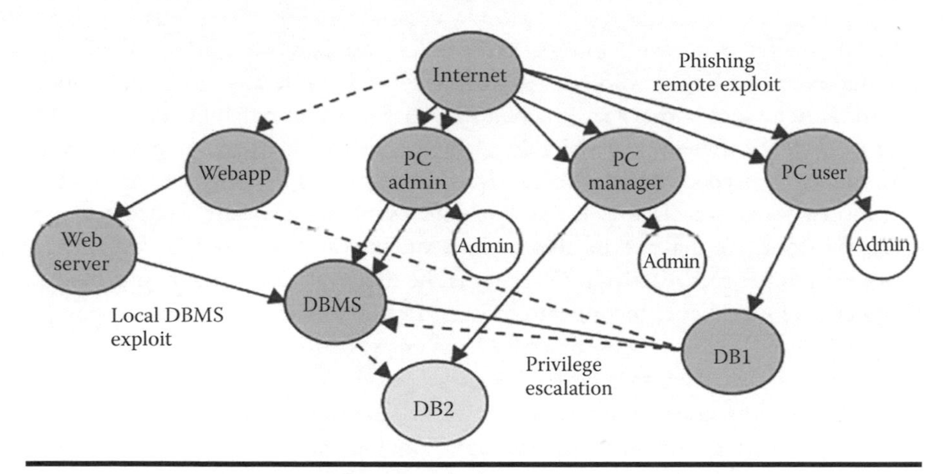

Figure 2.2 Network attack graph.

One element of the complexity of cyberspace is identifying the problem to be solved or the issue to be confronted. Another element is the underlying competitive nature of the attack-defender dynamic. Hackers and malicious systems are pitted against defenders of information and systems performance. Attacking elements often try to hide their true identity. Defenders try to hide or change their patterns so attackers cannot find weaknesses. This game theoretic setting takes modeling to new heights in what-if, cause-effect, and forensics questions. Cyber problem solving is an unstructured process that often requires high-dimensional data, nonlinear models, and dynamic modification to adapt to the constantly changing situations.

An analogy of cyber systems to biological systems can be helpful. Monocultures are efficient but vulnerable because uniformity and patterns create a form of weakness to non-normal conditions. Polycultures and diversity are inefficient, but usually robust to a changing environment and therefore survivable (Shah, 2014). The result is that diversity creates a form of strength. The ultra-efficiency of uniform order can produce fragility. This idea of adding randomness and diversity is not intuitive to modelers who have worked in other domains. Fortunately, despite our intent to make the Internet more uniform and efficient, it is not. The Internet works because of its inherent diversity and randomness. Yet we continue to follow our intuition and design our networks and systems primarily for efficiency. The result can be super-efficient networks that are often rigid and brittle. Even today, we revert to old, yet unhelpful, habits. When things go wrong in our cyber systems, we react by enforcing rigid discipline and control that destroys diversity and ultimately hurts the robustness and resilience. So, when weaknesses are found, cyber scientists may need to design in more intentional randomness into the network. Modelers use the system's diversity for improved survivability at the cost of efficiency. Randomness means that no one (not even the designer or builder) has precise control, but overall performance will still be higher than highly patterned, overprogrammed, inflexible, and eventually broken systems.

What makes a network robust, survivable, and hard to kill paradoxically also makes it inefficient, difficult to manage, and vulnerable to penetration. Evolutionary biology shows that inherent diversity provides reliability at a price of some inefficiency. Evolutionary biology also teaches that change (adaptation and randomness) is needed in order to survive. Today's cyber systems are vulnerable and unpredictable—a place where actions and events happen fast. So, to survive on the network, you have to be able to react quickly and effectively—sometimes proactively, sometimes reactively. Diversity is the model attribute that best provides the potential for resilience to vulnerabilities and yet maintains the agility to change fast. One natural way to create diversity in cyber systems is through randomness (explicitly designed random processes). Nature provides diversity in its DNA and cells; cyber scientists need to build diversity and randomness into their systems.

Cyber modeling enhanced by computational game theory and simulation enables war gaming of the basic elements of the cyber competition. These games and simulations are used to test capabilities, probe for vulnerabilities, fix performance degradation, and exercise the cyber systems. These are the research tools needed to enhance cyber security. Artificial intelligence techniques like machine learning and reinforcement learning are also elements in the models of the cyber framework. These faster-paced simulations will test the more advanced techniques of attacking and defending.

At the intersection of game theory and cyber security, models for static physical security games, such as Stackelberg Security Games, where defenders set a strategy and attackers surveil and attack, can be modified with more active defenders (Delle Fave et al., 2015; Shieh et al., 2016; Sinha et al., 2015). This framework gives a start for designing a fluid cyber situation with a defender trying to protect weighted targets, such as data servers, high-valued communication nodes, and physical access links. Attackers use probes, malware, exfiltration, and spoofing to attack the network. Defenders use honey pots, auditing, access control, detectors, and randomization of allocations and resources. These fluid actions are simulated so measures of the network and the strategies can be monitored and modeled for forensic analysis.

Researchers also experiment with smart cyber information systems with various security systems and test the use of randomness on performance and security measures. These tests can validate the framework and value of randomness in security. The goal of this computational approach is to develop measures of efficiency. The algorithms will defend or attack networked systems, predict and defend against attacks, respond to problems, use flexible and random designs, protect information, and monitor performance.

This information, coupled with machine learning techniques, could create autonomous defensive positions within cyberspace that, as attackers' techniques change, could modify a network's defenses without the intervention of a human. Being able to operate and respond at machine speed to threats is what military cyber forces require and what NS and DA can provide.

Ultimately, cyber science as a form of information science is different from the traditional sciences in terms of its competitive nature and highly structured network. Network models, Big Data analysis, game theory, and artificial intelligence are applied components of the digital cyber world. Computational game theory on a network is an important component of understanding the essence and dynamics of cyber science. As in many competitive security applications, the avoidance of operational and structural patterns through the inclusion of randomness and diversity are fundamental elements of cyber science. Therefore, this makes complexity an inherent element in cyber problems and their analysis. Cyberspace itself is the combination of many digital-related components that store, process, secure, protect, transmit, and use information. Network models establish a framework that incorporates the technical aspects of cyber operations, along with many human-based disciplines such as philosophy, ethics, law, psychology, policy, and economics that contribute to cyber analytics. Cyber science offers concepts and tools to understand complex security issues. In every domain and subject, there are differences in how data analytics and network modeling are used. Network models represent the connected elements and capture the dynamic of the attacker–defender interface. The cyber components that we study through NS and DA include:

- Authentication procedures
- Connections
- Operating systems
- Protocols
- Topology

There is an evolving future for data analytics and network modeling in cyber science. Cyberspace does require new, original ways of thinking and building models for the tasks that are part of this rapidly changing science. It is not often that the way ahead in a science is to implement randomness, embrace diversity, accept inefficiency, tolerate complexity, and thrive through interdisciplinary study. Cyber science in this form will be a challenge for our analytic community to accept, understand, and develop.

In a recent cyber-related application, a student at the United States Military Academy is using NS principles to explore whether submarine communications cables could pose a national security threat. Ninety-seven percent of the world's telecommunications traffic traverses these cables (shown in Figure 2.3), which are privately owned (Burnett et al., 2014). By creating a multilayer graph, he is exploring the interactions between the physical locations of the cable terminations, the multinational ownership conglomerations of the cables, and alliances between nation-states. History is full of examples of nation-states that tap these cables to steal information or deny access. NS will facilitate the simulation of a country denying access (removal of nodes and edges) and the second-/third-order effects to the ability to pass military operational traffic.

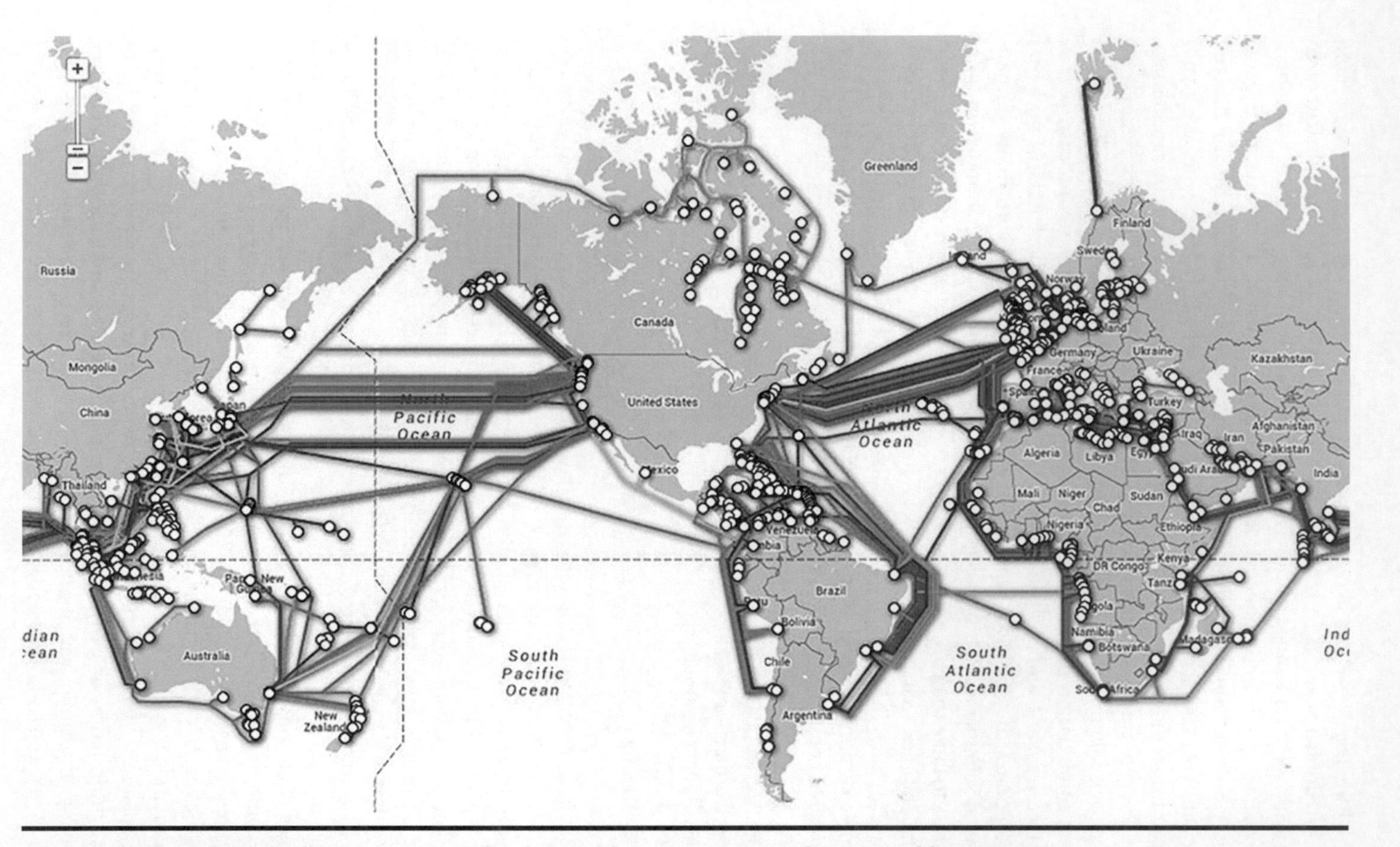

Figure 2.3 Snapshot of submarine communications cables. (From www.submarinecablemap.com.)

In another cyber-related application, analysts collect and study Twitter data to analyze the onset and evolution of social movements (Brantly, 2017; Korolov et al., 2016). They developed a framework to identify protest mobilization in social media and assess the likelihood of the protest occurring for a specified cause at given geographic location. Using machine learning, natural language processing, sentiment analysis, and influence network metrics, they study the tipping point and characteristic sentiment in the cyber domain to a kinetic action. By considering attempts to incite the protest or to influence public opinion via social media as a cyber-attack, they design methods to identify factors and mechanisms of protest development to suggest interventions. Future work will focus on identification of the antecedents of protest.

Complexity Science

The pace of discovery and progress in mathematics, science, society, and education continues to advance as the information, techniques, issues, and important questions shift (West, 2015). Science may be an evolutionary process, but the ways humans perform it, utilize it, and understand it can be revolutionary (deSolla Price, 1986; Weaver, 1948). Elements of this shift include: moving from little science (single investigators) to big science (large institutions supervising and controlling large groups) to team science (multidisciplinary, multi-organization, multiskilled, multination all-star teams of scientists); the increased role of complexity; connections in understanding the informational and networked nature of science; and the human utilization of science within the context of society (Holtz, 2015; Ledford, 2015; Scientific American Editors, 2015). Modern science through NS and DA is building a collective power that is more creative, more original, and more effective than the single disciplinary perspectives of the past. Some of this shift is attributed to the development of fractals and fractional mathematics (fractional calculus with fractional networking and fractional statistics) playing fundamental roles. Scientists hope to contribute to science's system of shared knowledge, appropriate levels of abstraction, and an enlightened human context that enables science to engage in the most compelling issues of society. It is no accident that this movement toward the integration of science coincides with a globalization of efforts and the proliferation of networks.

In our modeling and analysis efforts, we build models to understand the empirical differences in the shift of science. For example, we compared the collaboration networks for two exemplars of the shift (Paul Erdős [circa 1935–1995] and László Barabási [circa 1995–2015]). The networks we analyzed were the coauthor networks of the primary with the primary excluded. These are networks of co authorships of the coauthors, or the two degrees of separation network of the primary's coauthors. The differences in these networks indicate significant and dramatic change is taking place in scientific research, demonstrating the changing methods of science.

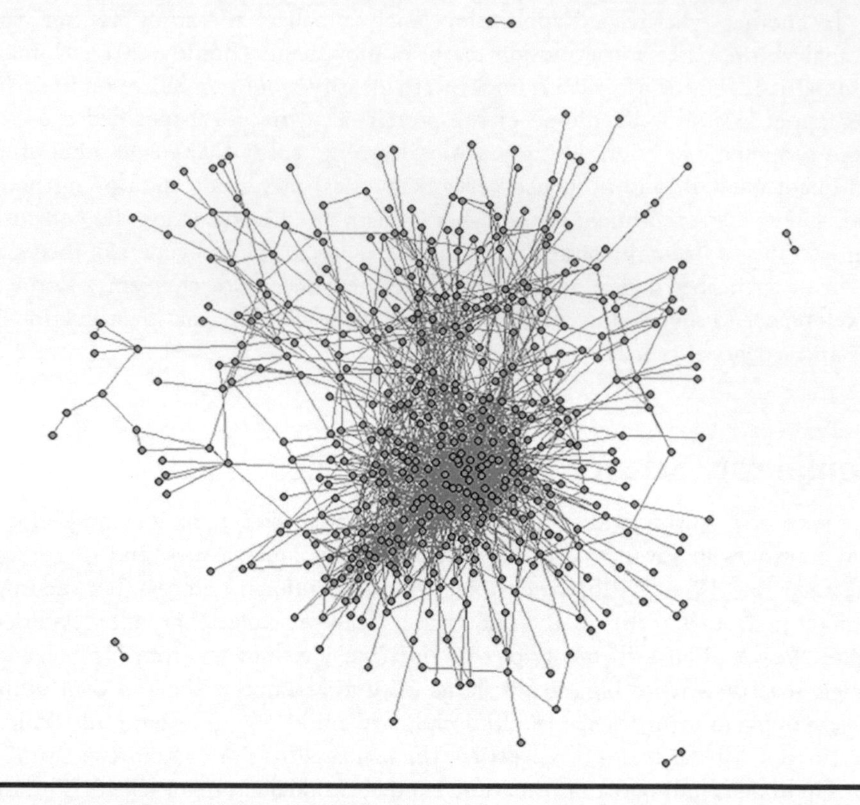

Figure 2.4 Erdős collaboration network.

For example, the Erdős network (a small science network) shown in Figure 2.4 has a much different structure core and peripheral structure than the Barabási network (team science), as shown in Figure 2.5. The Erdős network is also much less dense because of the smaller groups (often one person) that Erdős worked with.

The degree distribution graphs also show these same fundamental differences. The Erdős collaboration network distribution is shown in Figure 2.6 and the Barabási network distribution in Figure 2.7. The Barabási network distribution contains many more large-degree nodes enabling and demonstrating the team science approach to his research and problem-solving efforts.

The data summary for the two networks is provided in Table 2.1. The Erdős co authors have a mean degree of 6.76, whereas the Barabási co authors are much more connected with a mean degree of 28.4 and much more clustered into dense subnetworks with a much higher cluster coefficient.

DA and NS in the context of team science help us to understand the next layers of modern science that are the result of complexity (often created by unseen networks

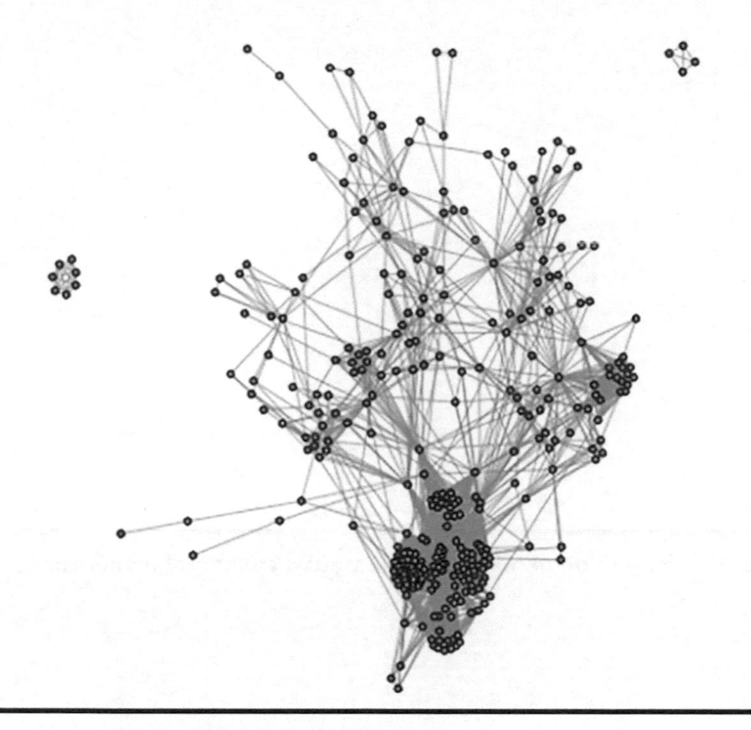

Figure 2.5 Barabási collaboration network.

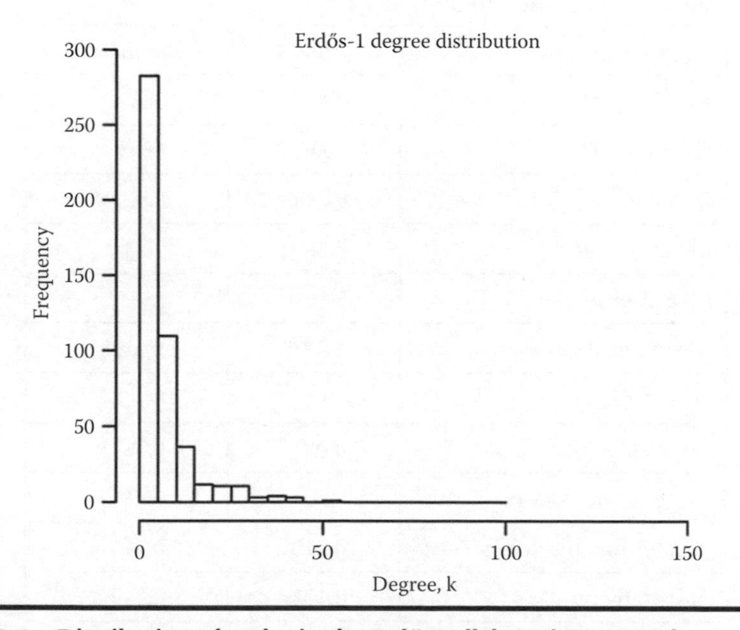

Figure 2.6 Distribution of nodes in the Erdős collaboration network.

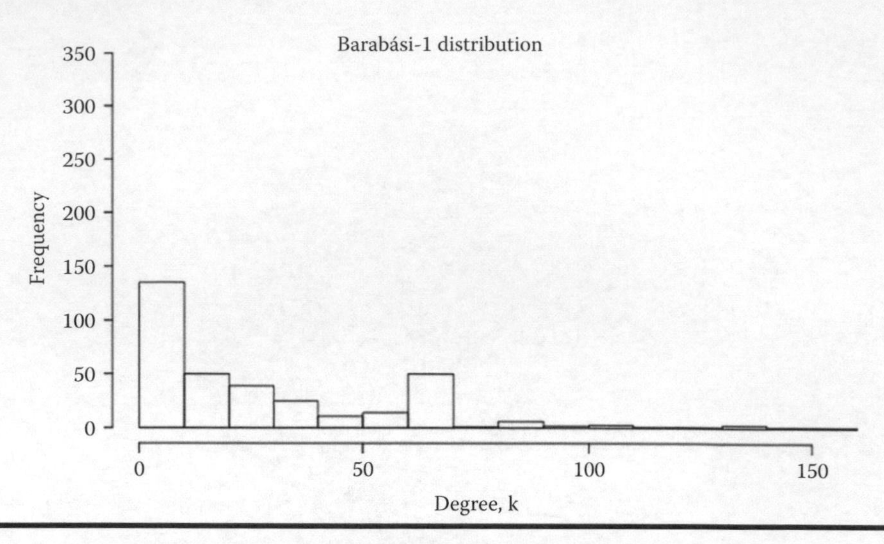

Figure 2.7 Distribution of nodes in the Barabási collaboration network.

Table 2.1 Network Measures for Erdős and Barabási Collaboration Networks

Summary Statistic	*Erdős-1*	*Barabási-1*
Node count	474	343
Edge count	1604	4883
Density	0.0143	0.08315
Mean degree	6.76	28.4
Max degree	51	154
Mean distance	3.8435	2.9268
Diameter	10	8
Centralization	0.093	0.3671
Largest component	98%	96%
Clustering coefficient	0.219	0.714
Assortativity	0.177	0.249

or dark portions of networks) that is still not well understood. Complex phenomena remain ill-understood because the traditional methods of data and mathematical analysis are insufficient to overcome that barrier of complexity. One way to advance that understanding is with better clustering of datasets.

Clustering Methods and Metrics

Clustering is a DA technique that can be applied to any collection of data with the goal of organizing them into clusters—subsets that have similar characteristics. The underlying principle is to define a distance measure between pairs of records and then partition the records for which the distances between pairs of records in the same cluster are small, and the distances between pairs of records in different clusters is large (Bertsimas et al., 2016).

In many applications, distance is determined by differences of values in a set of selected attributes. This permits weighted attributes reflecting the relative importance of the difference in each selected attribute. Part of the art of using this method is to define distances in a way that the resulting clusters have the desired property—data in each cluster have similar values for the selected attributes and data in different clusters have different values. Often clustering is the first step in data analysis. There are no universally accepted metrics used to evaluate clustering methods. There may be different objectives for different uses of clustering. In some cases, the user wants to predefine the number of clusters to be formed; in other cases, this is left unspecified. In some cases, it is desired that every record belongs to one set in the cluster. In other applications, not all records need to belong to a cluster, and some records may be permitted to belong to more than one cluster.

We measure the effectiveness of a clustering method by computing a penalty p for each pair of records that are close to each other and in different clusters, and a penalty q for each pair of records different from each other but in the same cluster. The size of the penalty for a pair of records depends on how similar or different the records are according to the defined distance function. An optimal clustering method corresponding to a specified metric is one for which the sum of the penalties, over all pairs of records, is minimized. This methodology and algorithm are similar to the linear discriminant used to find attributes that separate the dataset into different kinds of objects where the attribute being considered is geographic location.

Conclusions

Network science and data analytics are playing important roles in developing cyber science and complexity science. Our work hopes to build a foundation for further development and the application of these concepts and methods to data-related problems in military-relevant areas, such as cyber and complex science.

References

Arney, C. (2016). Cyber modeling. *UMAP Journal*, 37, 93–97.

Arney, C. and Coronges, K. (2015). Categorical framework for complex organizational networks: Understanding the effects of types, size, layers, dynamics and dimensions. In G. Mangioni, F. Simini, S. Uzzo, and D. Wang (Eds.), *Complex Networks VI*, pp. 191–200. doi:10.1007/978-3-319-16112-9_19.

Bertsimas, D., O'Hair, A., and Pulleyblank, W. (2016). *The Analytics Edge*. Belmont, MA: Dynamic Ideas.

Brandes, U., Robins, G., McCranie, A., and Wasserman, S. (2013). What is network science? *Network Science*, 1, 1–15.

Brantly, A. (2017). Innovation and adaptation in jihadist digital security. *Survival: Global Politics and Strategy*, 59, 79–102.

Burnett, D., Beckman, R., and Davenport, T. (2014). *Submarine Cables: The Handbook of Law and Policy*. Boston, MA: Martinus Nijhoff Publishers.

Carter, K., Riordan, J., and Okhravi, H. (2014). A game theoretic approach to strategy determination for dynamic platform defenses. In *Proceedings of the First ACM Workshop on Moving Target Defense*, New York: ACM, pp. 21–30.

Delle Fave, F.M., Shieh, E., Jain, M., Jiang, A., Rosoff, H., Tambe, M., and Sullivan, J. (2015). Efficient solutions for joint activity based security games: Fast algorithms, results and a field experiment on a transit system. *Journal of Autonomous Agents and Multiagent Systems*, 29, 787–820.

Department of Defense (2013). *Joint Publication 3-12 (R) Cyberspace Operations*. Washington, DC: US Department of Defense.

Department of Defense (2015). *The DOD Cyber Strategy*. Washington, DC: US Department of Defense.

deSolla Price, D.J. (1986). *Little Science, Big Science… and Beyond*. New York: Columbia University Press.

Holtz, R. (2015). How many scientists does it take to write a paper? Apparently, thousands. *The Wall Street Journal*, August 9.

Ingols, K., Chu, M., Lippmann, R., Webster, S., and Boyer, S. (2009). Modeling modern network attacks and countermeasures using attack graphs. In *Proceedings of the 2009 Annual Computer Security Applications Conference*, Honolulu, HI, pp. 117–126.

Korolov, R., Lu, D., Wang, J., Zhou, G., Bonial, C., Voss, C., Kaplan, L., Wallace, W., Han, J., and Ji, H. (2016). On predicting social unrest using social media. In *ASONAM 2016*, San Francisco, CA, August 18–21, pp. 89–95.

Ledford, H. (2015). How to solve the world's biggest problems. *Nature*, September 21.

Mayhew, M., Atighetchi, M., Adler, A., and Greenstadt, R. (2015). Use of machine learning in big data analytics for insider threat detection. In *Proceedings of the MILCOM 2015 – 2015 IEEE Military Communications Conference*, Tampa, FL, pp. 915–922.

National Academies of Sciences, Engineering, and Medicine (2017). *Strengthening Data Science Methods for Department of Defense Personnel and Readiness Missions*. Washington, DC: National Academies Press. doi:10.17226/23670.

National Research Council (2005). *Network Science*. Washington, DC: National Academies Press.

Paxton, N., Moskowitz, I.S., Russell, S., and Hyden, P. (2014). Developing a network science based approach to cyber incident analysis. *Paper presented at NATO IST-122 Symposium on Cyber Security*, Talinn, Estonia.

Roehner, B.M. (2007). *Driving Forces in Physical, Biological, and Socio-economic Phenomena: Network Science Investigation of Social Bonds and Interactions.* Cambridge, UK: Cambridge University Press.

Scientific American Editors (2015). State of the world's science 2015: Big science, big challenges. *Scientific American*, 313, 34–35. doi:10.1038/scientificamerican1015-34.

Shah, A. (2014). Why is biodiversity important? Who cares? *Global Issues*, January 19. Accessed August 9, 2017. http://www.globalissues.org/article/170/why-is-biodiversity-important-who-cares.

Shieh, E., Jiang, A., Yadav, A., Varakantham, P., and Tambe, M. (2016). Extended study on addressing defender teamwork while accounting for uncertainty in attacker defender games using iterative dec-MDPs. *Multi-Agent and Grid Systems*, 11, 189–226.

Sinha, A., Nguyen, T., Kar, D., Brown, M., Tambe, M., and Jiang, A. (2015). From physical security to cyber security. *Journal of Cybersecurity*, 1, 19–35. doi:10.1093/cybsec/tyv007.

Weaver, W. (1948). Science and complexity. *American Scientist*, 36, 536–544.

West, B. (2015). *Fractional Calculus View of Complexity: Tomorrow's Science*. Boca Raton, FL: CRC Press.

Chapter 3

Context for Maritime Situation Awareness

Anne-Laure Jousselme and Karna Bryan

Contents

Introduction

Military shortfalls in information processing, exploitation, and dissemination will be partially addressed by current industrial and academic research in areas like Artificial Intelligence and Big Data. The Big Data challenge is often recognized in terms of Volume, Velocity, Variety, and Veracity. This highlights the need to process an always increasing amount of data (challenges with volume) within strong time constraints (challenges with velocity) originating from diverse sources with different formats, requiring some alignment in time, space, or semantics (challenges with variety) and with many imperfections such as uncertainty, imprecision, errors, and conflict (challenges with veracity). In addition to these challenges in data processing, there is also an underlying semantic challenge of providing valid interpretations and meanings of the information to the decision maker.

Moving toward improved synergy between humans and machines, research on the automation of data processing should be accompanied with research on the automation of reasoning. The artificial intelligence (AI) community, and more recently that of Information Fusion, address challenges in automation and information processing. While "[d]ata fusion refers to the combination of data from multiple sensors (either of the same or different types), [...] information fusion refers to the combination of data and information from sensors, human reports, databases, as well as a broad range of contextual data."* Contextual information is a fundamental aspect of the reasoning process, and the notion of context has become a key component in many studies of these fields. In Snidaro et al. (2015), the authors survey the recent works on context-based information-fusion systems, mapping them to the joint directors of laboratories (JDL) levels (Llinas et al., 2004). They discuss the use of context in the classical lower-level tasks of sensor characterization, signal fusion, data association, filtering, tracking, and classification, and consider as well the higher-level tasks of knowledge representation, situation assessment, decision making, intent assessment, and process refinement. In relation to natural language processing, context is also used in hard and soft fusion (Jenkins et al., 2005) to incorporate soft information (i.e., from human sources) with hard information (i.e., from sensors) using fuzzy membership functions to define the semantics of vague concepts. Indeed, with the growth of social media data processing, the notion of context is instrumental to natural language processing techniques, where words and sentences often need some contextual setting to be properly understood.

Effective maritime situation awareness (MSA) relies on high-level information fusion tasks that benefit from a formal approach to context-based reasoning. Indeed, MSA requires not only detecting, tracking, and classifying vessels but also detecting, classifying, and predicting any abnormal behavior, which includes detecting relationships between vessels and vessels' behaviors. MSA operations occur in both peacetime and times of conflict, and involve the understanding of the patterns of life of all entities operating in and influencing the maritime domain. Situational awareness includes the development of context by military operators. This context applies to various aspects of a processing chain with the layers of the chain building on each other. At the core of this task is the compilation of a maritime picture (Lane et al., 2010; Laxhammar, 2008), which involves extracting relevant contextual information (for instance, maritime routes or loitering areas [Pallotta et al., 2013]) but also monitoring the real-time maritime traffic. The use of a set of sensors mixing cooperative self-identification systems, such as

* www.buffalo.edu/cmif/center/what-is-MIF.html.

the automatic identification system (AIS), and noncooperative systems, such as coastal radars or satellite imagery, provides the necessary complementarity and redundancy of information to overcome the possible (and quite common) spoofing of AIS signals while increasing the clarity and accuracy of the maritime picture. In many cases, intelligence information is also of great help to refine and guide the search in the huge amount of data to be processed, filtered, and analyzed, and can be contextual information for some MSA problems.

In this chapter, we discuss the relative notion of context through the problem of Maritime Situation Awareness. In the section "Maritime Situation Awareness," we introduce the MSA problem, exemplifying the three embedded problem-solving and associated contexts of route extraction, maritime anomaly detection, and threat assessment. In the section "Context-Based Maritime Situation Awareness," we introduce some definitions of context and discuss the impact and use of context from an information fusion perspective on the MSA problem. We revisit the MSA problems and present a scheme with embedded problems constrained by different contexts. We conclude in the section "Conclusions and Future Work" and discuss future works.

Maritime Situation Awareness

In order to present an overall and structured vision of the different contextual dimensions within an MSA perspective, this section presents three subproblems of MSA that are: route extraction ("Route Extraction"), maritime anomaly detection ("Maritime Anomaly Detection"), and threat assessment ("Threat Assessment [Identification]"). We highlight that context is a relative notion on these three embedded subproblems.

Route Extraction

The analysis of historical, spatiotemporal traffic data streams provides baseline information on Patterns-of-Life (PoL). The AIS cooperative self-reporting system provides a rich data stream that can be used to extract PoL for maritime vessel traffic. Machine learning algorithms are one way to derive PoL, as they provide a suitable degree of automation and efficiency in detecting and characterizing inconsistencies, anomalies, and ambiguities, and ultimately transform this information into usable and actionable knowledge (Pallotta and Jousselme, 2015). The Traffic Extraction and Anomaly Detection (TREAD) tool (Pallotta et al., 2013) developed at the Centre for Maritime Research and Experimentation (CMRE) implements an unsupervised classification approach, deriving a dictionary of the maritime traffic routes using spatiotemporal data streams from terrestrial and/or satellite AIS receivers. Figure 3.1 displays the overview of the TREAD process, from raw AIS data trajectories to route definitions.

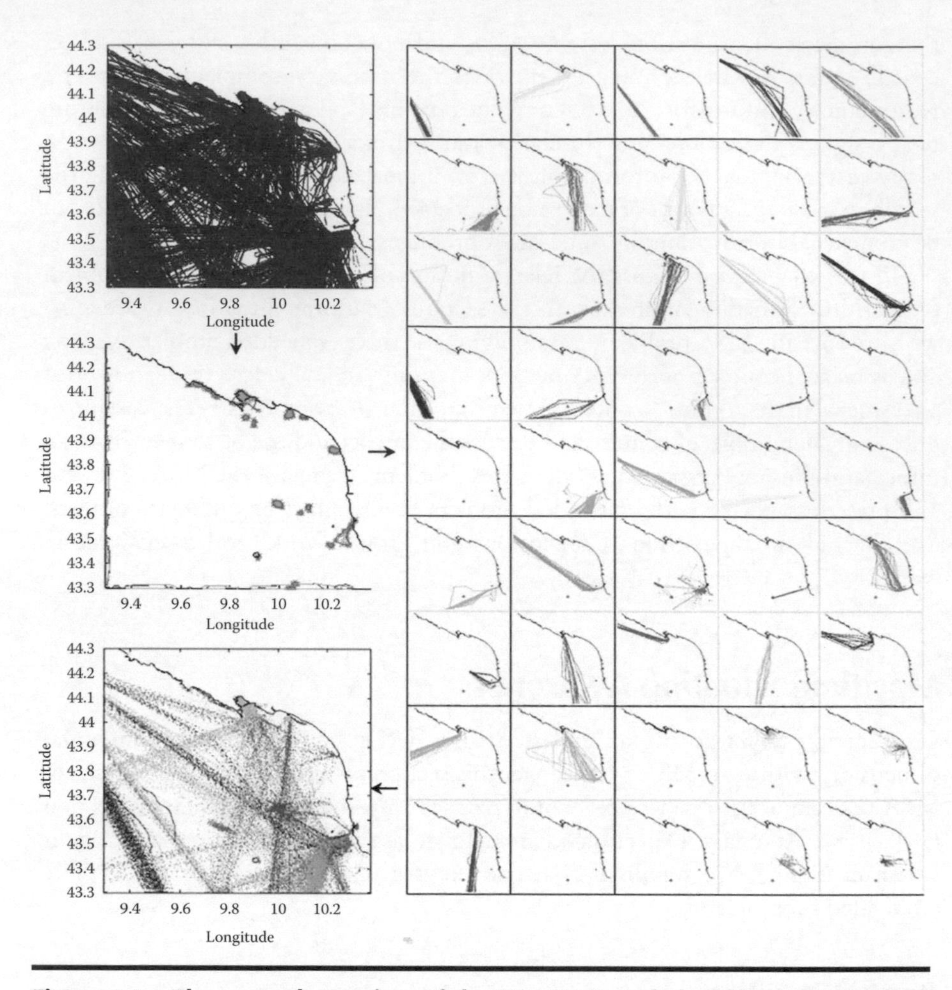

Figure 3.1 The general overview of the TREAD Knowledge Discovery Process: The raw AIS data stream is processed, vessel movements are clustered, and as a result, the discovered traffic route system is organized as a dictionary of motion models. (From Pallotta, G. et al., *Entropy*, 5, 2218–2245, 2013.)

The synthesis of the activity at sea summarizes the maritime traffic over a given period of time and a given area and is referred to as the maritime routes. A TREAD route is then defined by a starting point and an ending point, together with a subset of waypoints, describing a path on a portion of the sea. Starting and ending points for the routes are stationary areas. Stationary areas can either be coastal areas such as ports, including island or offshore ports, or open-sea areas, such as fishing areas. They can also be entry or exit areas within the area of interest. The TREAD algorithm clusters individual vessel contacts, each cluster then corresponding to a route.

The problem of Route Extraction (RE) can be formalized as:

$$\Psi_{RE} : \mathcal{X}_{P,\Delta t} \rightarrow \mathcal{R} \tag{3.1}$$

$$\mathbf{x} \mapsto R \tag{3.2}$$

where $\mathcal{X}_{P,\Delta t}$ is the measurement of space and by extension, the dataset of positional data streams for a set of vessels over a given period of time Δt; $\mathbf{x}$ is a trajectory; R is the label assigned to $\mathbf{x}$ belonging the $\mathcal{R}$, the finite set of routes. The route extraction problem aims at assigning each contact $\mathbf{x}$ to a class R such that $\Psi_{RE}(\mathbf{x}) = R$ is the route label associated with $\mathbf{x}$, and R is the corresponding route. Within an unsupervised classification method (clustering), such as the one used by TREAD, the decision space is not known in advance and the set of labels emerges from the data, thus building the decision space. While only temporal streams of positional information are processed to extract the set of maritime routes, the routes are further characterized by additional attributes representing the traffic of the vessels composing it, such as the speed, the type, and the heading. The associated uncertainty characterization of the route along these attributes can be more or less complex, ranging from average values only, added variance parameters, complete probability distributions, or sets of distributions. Table 3.1 lists an excerpt of the dictionary of routes together with some possible uncertainty representations (Jousselme and Pallotta, 2015).

Table 3.1 Dictionary of Routes and Associated Uncertainty Representation

Route	*Synthetic Route*	*Traffic Information*			
	Position	*Course*	*Speed*	*Length*	*Type*
	$\{(p_i^{(k)}, \theta_i^{(k)})\}_{i=1}^{N} \pm w^{(k)}$	$\bar{\theta} \pm 2\sigma_\theta$	$\bar{s} \pm 2\sigma_s$	$[l_{\min}; l_{\max}]$	P over $\mathcal{X}_T$
R_1	$\{WP\}^{(2)} \pm 5$ km	$210 \pm 25^\circ$	$\mathcal{N}(11,2)$	[80;200]	[10000]
R_2	$\{WP\}^{(2)} \pm 2$ km	$245 \pm 30^\circ$	$\mathcal{N}(12,3)$	[20;170]	[0.43 0.43 0.05 0.05 0.04]
R_3	$\{WP\}^{(3)} \pm 1$ km	$280 \pm 30^\circ$	[2;16]	[20;290]	[0.5 0.5 0 0 0]
R_4	$\{WP\}^{(4)} \pm 4$ km	$185 \pm 10^\circ$	[10;16]	[20;230]	[0.75 0 0 0 0.25]
R_5	$\{WP\}^{(5)} \pm 1.5$ km	$325 \pm 20^\circ$	$(\mathcal{N}(12,2); \mathcal{N}(19,2))$	[110; 200]	[0.67 0.11 0 0.11 0.11]

For instance, the speed attribute for Route R_1 is defined by the tuple $(\bar{s}_1; \sigma_s^1)$ representing the mean and variance of speed values estimated on the training dataset of trajectories used to build R_1 of a possible normal distribution law. Another representation could be $(\bar{s}_1; \pm\sigma_s^1)$ defining an interval of speed values for R_1.

The maritime traffic (and thus the set of routes extracted) may be influenced by the meteorological conditions (i.e., some areas may be avoided due to specific bad weather), seasonality, economical conditions (i.e., ships may decide or change their destination based on the current stock market linked to their cargo), or even areas of conflict that they would like to avoid. Consequently, the route extraction process would benefit from considering contextual information. For instance:

- The location of specific geographically defined areas such as channels, restricted areas, fishing areas, borders, harbors (fishing, recreational, etc.), shipping lanes, ferry lanes, military, or liquefied natural gas (LNG) anchorage areas can assist by excluding some zones from the route trajectory.
- A port index with geolocalization would help to refine the starting and ending points of the routes.
- The economical situation relative to the objective or vessel of interest for the period when the data were collected would favor some trajectories.
- The period of the year, such as the season, or real-time meteorological conditions could eliminate some outliers in the route construction.
- The user's goals when extracting the routes could define a finer or coarser granularity of the clusters to be built.

In addition to showing how the route extraction process could be improved and refined by contextual information, a further observation can be made about context: This synthetic information, provided as the set of routes together with its associated uncertainty, characterizes part of the context for another problem-solving situation, the detection of anomalies at sea. Thus, the decision variable X_R becomes a context variable for another problem.

Maritime Anomaly Detection

While the extraction of PoL mainly relies on the statistical processing of large amount of data, generally using a single source of information, namely the AIS receiver, the use of several complementary sources of information is crucial for the anomaly detection task. Indeed, although the deviation from normalcy (i.e., pre-extracted PoL) can rely on a single source of information, it appears that other situational indicators may be revealed only by correlating the output of several sources, including cooperative and noncooperative ones. For instance, fusing the synthetic aperture radar (SAR) imagery with the AIS may help in detecting purposefully incorrect self-reported AIS information about the type of the vessel.

Let us thus consider a vessel of interest v among a set $\mathcal{O}$ of objects of interest ($v \in \mathcal{O}$) observed by a series of sources $\mathcal{S} = \{s_1, \ldots, s_n\}$ possibly being of different natures, such as a coastal radar and its associated tracker (s_1), a SAR image with associated automatic target recognition (ATR) algorithm (s_2), a human analyst (s_3), a visible camera operated by a human analyst (s_4), AIS information sent by the vessel itself (s_5), or other intelligence sources. The association problem (i.e., the task of associating a piece of information to an object v), although critical, is not considered here, and it is assumed that all the pieces of information are already associated to v.

Let $\mathcal{A}$ = {***Position,***Heading,***Speed,***Length,***Type} be the set of attributes of interest to be observed by the set of sources, and let $\mathcal{X}$ be the corresponding observation space:

$$\mathcal{X} = \mathcal{X}_P \times \mathcal{X}_H \times \mathcal{X}_S \times \mathcal{X}_L \times \mathcal{X}_T$$

Estimations of each attribute value a are possibly provided by more than one source, providing redundant information, while the same source may also provide information about different attributes, that is, complementary information. Let us denote $\mathbf{x}_t = \{\phi^{(s_1)}_{p,t}; \phi^{(s_2)}_{\theta,t}; \phi^{(s_3)}_{s,t}; \phi^{(s_4)}_{l,t}; \phi^{(s_5)}_{T,t}\}$, a set of heterogeneous observations jointly provided by the set of sources about the attributes in $\mathcal{A}$. This notation covers the general case where sources are able to provide some uncertainty about their statement and thus ϕ denotes each source's statement either as a single measurement, a probability vector, a natural language declaration, or another quantity. In the specific case of precise and certain measurements from the sources, $\mathbf{x}_t$ is a vector of $\mathcal{X}$. The maritime anomaly detection (MAD) problem aims at establishing a mapping:

$$\Psi_{AD} : \mathcal{X} \rightarrow \Omega_{AD} \tag{3.3}$$

$$\mathbf{x} \mapsto \omega \tag{3.4}$$

from the observation space $\mathcal{X}$ to the decision space Ω such that $\omega = \Psi_{AD}(\mathbf{x})$ is the anomaly label assigned to a vessel v under observation, itself represented by the observation vector $\mathbf{x}$ (at time t). The decision space Ω_{AD} could be built, for instance, from a series of anomalies of interest such as:

- ω_0: "The vessel is physically off-route."
- ω_1: "The vessel is physically on-route with too high of speed."
- ω_2: "The vessel is physically on-route traveling in the reverse direction."
- ω_3: "The vessel is physically on-route with incompatible vessel type."

In this case, we note that Ω is neither exhaustive (v can follow other abnormal behaviors) nor exclusive (v can follow more than a single abnormal behavior).

The consideration of these two hypotheses in the problem model is crucial as, for instance, probability theory requires that the universe of discourse is a both exhaustive and exclusive set of hypotheses.

For this problem of anomaly detection, the context provides not only the normalcy conditions (or what is expected) based on historical PoL, but also current information about anything that may impact (and thus explain) the actual behavior of the vessel. The contextual elements of interest for anomaly detection include:

- The set of extracted routes (see the section "Route Extraction"), $\mathcal{R} = \{R_1, \ldots, R_K\}$ for the given area over the period of time corresponding to the actual period, therefore providing the set of routes the vessels are expected to follow.
- The site, including harbor zone characteristics such as water depths, channels, restricted areas, fishing areas, borders, harbors (fishing, recreational, etc.), shipping lanes, ferry lanes, military, or liquefied natural gas (LNG) anchorage areas that would explain any deviation from a normal path (expected trajectory).
- The geopolitical situation or the global security alert state, which directly impacts the rules and decisions.
- The meteorological conditions (sea state, weather, etc.).
- The traffic density, which would give expected speed values for the vessel.
- Prior information about the situation, for instance, just before the triggering event.
- Sources and devices providing the information, including their quality.
- The user's goals in detecting anomalies.

We note that the notion of "anomaly" is itself contextual, as it depends on the user's information needs as well as on the particular circumstances of the situation. What is abnormal to a given user is what does not match his/her expectations at a specific time. The expectations could be defined by some normalcy, dependent itself on a specific context.

Threat Assessment (Identification)

Detecting anomalies of interest is typically not an end in itself, as anomaly detection is often driven by the higher-level task of threat assessment. In threat assessment, the focus is the vessel's intent together with the possible consequences of an undesired event. The threat could come from any suspect activity, for example, a terrorist attack, illegal fishing, smuggling, or illegal immigration, which would require an intervention based on an informed decision following a judicious risk assessment.

Target identification (TI) is presented in Bryant (2009) (citing other authors) as an element of combat identification (CID) together with situation awareness (SA) and tactics, techniques, and procedures (TTPs). CID is itself defined by Department of Defense (2000) as "[...] the process of attaining an accurate characterisation of entities in a combatant's area of responsibility to the extent that high-confidence, real-time application of tactical options and weapon resources can occur." While the recognition task mainly assigns a category (or class) to the vessel under observation, the threat assessment task aims at determining if the target is hostile or friendly. Although we easily understand how these two tasks are linked, especially in an asymmetrical threat context, any object may happen to be hostile, not only those in a particular category. Threat assessment thus relies as well on some behavior and intent assessment, some elements being provided by the anomaly detection task.

The measurement space for the threat assessment problem is built from variables corresponding to different attributes including some anomaly detection outputs, but also other classification information such as kinematic behavior, electromagnetic emission, IFF (Identify Friend of Foe) answers, origin, or nationality. The third edition of NATO STANAG 4162 "Identification Data Combination Process" (STANAG 4162, 2001) provides an exhaustive list of sources of information leading to the "identification" of an object.*

The identification task aims to build a mapping from the observation (or measurement) space to the decision space, for instance:

$$\Psi_{ID} : \mathcal{X} \rightarrow \Theta_{1241} \tag{3.5}$$

$$\mathbf{x} \mapsto \theta \tag{3.6}$$

where STANAG 1241 defines the list of standard identities (STANAG 1241, 2005):

$$\Theta_{1241} = \{\text{Unknown (U), Assumed Friend (AF), Friend (F)} \\ \text{Neutral (N), Suspect (S), Hostile (H)}\} \tag{3.7}$$

so that $\theta = \Psi_{ID}(\mathbf{x})$ is the NATO standard identity label assigned to the vessel v under observation, itself represented by the observation vector $\mathbf{x}$ (at time t).

While the level of automation for the recognition task can be higher, the identification task itself requires a closer interaction with the human: An information support system would at most suggest labels to the Identification Authority (STANAG 4162, 2001). The identification task does not only rely on sensors' measurements but also on contextual information and risk assessment. For instance, the same object may be assigned a Suspect ID label under a harbor protection level (HPL) of 2, while the exact same vessel with the exact same behavior would be assigned a label Assumed Friend under an HPL of 1.

* "[T]he term identification encompasses all the Military Operational Requirements related to identification: Allegiance, Civil/Military distinction, Platform, Platform Activity (or intent), Specific Type and Nationality" (STANAG 4162, 2001).

Context-Based Maritime Situation Awareness

In the following, we discuss why the notion of context should be considered in the design of future information fusion systems for supporting Maritime Situation Awareness. We discuss some categorization of context in relation to the famous JDL model of fusion, and provide a general scheme of context-based reasoning in fusion applications.

Why Use Context?

While a unique reference for a definition of context probably does not exist, the ideas conveyed by the different definitions converge. For Dey (2001), "Context is any information that can be used to characterise the situation of an entity. An entity is a person, place, or object that is considered relevant to the interaction between a user and an application, including the user and applications themselves." Brezillon (2003), Winograd (2001), and Bouramoul et al. (2001), from a rather philosophical perspective, state that "There is no context without context." For Brezillon (1999), "Context is what constrains a problem solving without intervening in it explicitly," and, "context is inseparable from its use." For Winograd (2001), "The context is a set of information. This set is structured, it is shared, it evolves and serves the interpretation."

We follow the conceptualization of context in AI, meaning that:

1. Context exists only if it is useful, thus its strong relationship with the notion of relevance.
2. The use of context depends on the purpose (Bouramoul et al., 2011; Brezillon, 1999).
3. Context acts as a filter to focus on and better scope problem solving.

Facing the huge volume of various information, which often lacks veracity, the operator does not only need the appropriate information with sufficient quality to make his/her decision, but also needs to understand its underlying meaning (its origin, how it has been obtained, processed, what was the context of its creation, etc.). For instance, it is of interest for the vessel traffic system (VTS) operator to understand how an "anomaly detector" came up with an alert: Which were the reference data? Which sources were processed? Was the information and associated uncertainty obtained in an objective or subjective manner? Did the process consider the sources' quality, and if so, how? Was the contextual information considered? What is the meaning of the numerical value of the uncertainty output? What was the underlying logical reasoning providing the answer?

It is therefore expected that an adequate problem model explicitly including context specification and characterization will provide:

1. The necessary simplicity to understanding the processes by a judicious filtering of information corresponding to the users' needs.
2. The suitable flexibility for the algorithm's implementation by separating the events of interest from their surrounding context.
3. The adequate uncertainty representation and processing considering the sources' quality and uncertainty's origin.
4. The expected human-system synergy for a better understanding of the system's outputs with associated explanations and simpler queries tuned to specific needs.

Categorization of Context and Fusion

Additionally, to the expected benefit of considering context to improve MSA, it is also acknowledged that the processing of a wide diversity of sources would improve the situation awareness of the user. On the one hand, the complexity of the process requires a categorization of the tasks to be performed, which leads in particular to the JDL functional model of information fusion (Llinas et al., 2004). On the other hand, the diversity of contexts and context-uses motivates the categorization of context, such as the work of Razzaque et al. (2005) who identified some conceptual contextual parameters for context categorization.

Table 3.2 lists the six categories of context—*user*, *physical context*, *network context*, *activity context*, *device context*, and *service context*—exemplified along three levels of the JDL model.

For instance, the user context of the VTS operator who needs to detect and filter out anomalies at sea (i.e., JDL level 2 task) would determine the level of granularity required for the anomalies, the set of anomalies to be detected, and so on. The physical context would be, in this case, the environmental conditions and the restricted areas, but also the critical infrastructure for the level 3 tasks where the impact needs to be assessed. The device context is the set of sensors available, their locations (which defines their coverage, i.e., the vessels about which they are able to provide information), their performance, and their lifetime.

Contextual Problem Solving

Problem solving involves the processing of information provided by hard sources (e.g., sensors, algorithms) and soft sources, provided by human-generated

Table 3.2 Examples of Contextual Information for the Six Categories of Razzaque et al. (2005) along Three JDL Levels

Context Category	*Semantics*	*JDL Level 1 (Entity Assessment)*	*JDL Level 2 (Situation Assessment)*	*JDL Level 3 (Impact Assessment)*
User context	Who?	Tracking analyst with the goal of detecting and tracking all targets	VTS operator with the goal of detecting anomalies	Watch Officer with the goal of identifying relevant, suspicious, or hostile events
Physical context	Where?	Coastlines, sea state	Anchorage areas, channels	Critical infrastructures
Network context	Where?	Network of radars	AIS emission and communication network	Hierarchical chain of command, other command entities, intelligence analysts
Activity context	What occurs, when?	Usual tracks or routes	Season, routes, patterns of communication with port authorities	Social media activity, recent piracy/terrorist attacks
Device context	What can be used?	Radars, cameras, satellites	AIS coverage	Availability and accessibility of sources
Service context	What can be obtained?	Location (lat, long), heading (degrees)	Anomalies of interest	Standard identities (e.g., STANAG 1241)

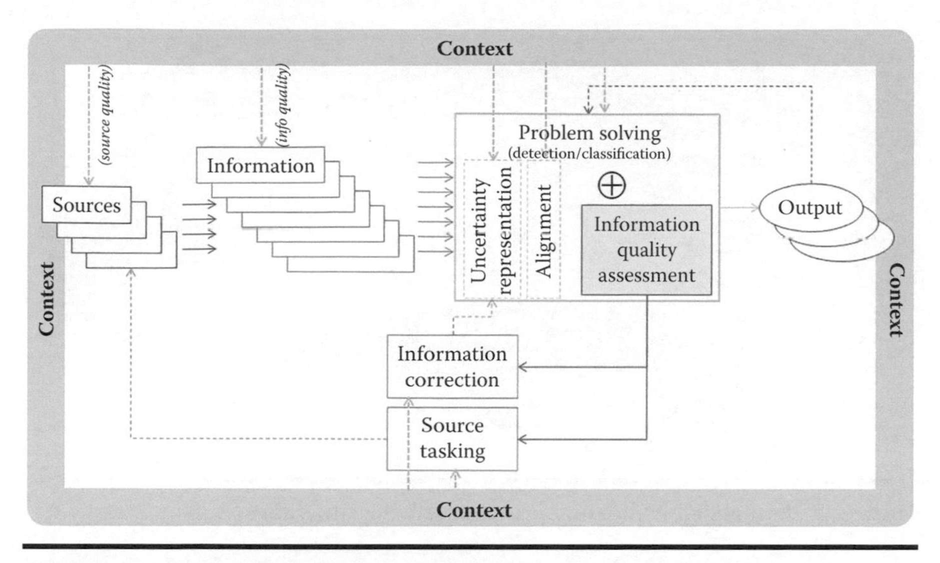

Figure 3.2 General overview of context-based information fusion.

information such as witnesses, social media, and intelligence reports. It outputs some inferred information with a possible feedback loop to either correct the information (e.g., based on conflict assessment and interpretation) or task the sources (e.g., modify or adapt the robot's pattern, reorient the radar). Figure 3.2 displays a general scheme of context-based reasoning with multiple sources of information, whose purpose is twofold: information correction and source-tasking for information gathering. These two corresponding tasks are two examples of the feedback (adaptive) loop recognized as the level 4 of the JDL model, where a system's performance is assessed and used to refine the process on the fly.

The information processing may be very complex but includes at least the following elements, which can possibly all be influenced by context (Llinas et al., 2016):

1. Some aspects of uncertainty representation or processing.
2. Some alignment processes as required for combination or fusion functions.
3. A combination/fusion function, putting in relation the different pieces of information, inferring new information.
4. An information quality-assessment process to detect conflicts or characterize the output quality for further use by source-tasking.

One way of formally considering context in fusion processes is to partition the set of variables in context and problem variables, as suggested in Steinberg and Bowman (2013). The three examples of MSA discussed in the previous section highlight the relative notions of context and problem-solving situations.

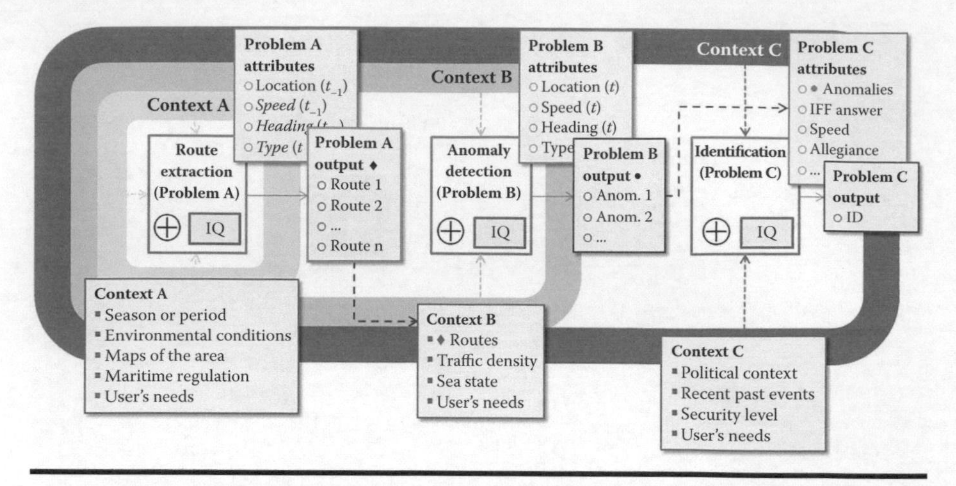

Figure 3.3 Embedded contextual problem-solving situations.

Figure 3.3 displays the three embedded problem-solving situations and their associated context as tuples (Problem A; Context A), (Problem B; Context B), and (Problem C; Context C). For instance, the route extraction problem is seen as contextual information for the maritime anomaly detection problem, which itself provides both situational and contextual information to the threat assessment problem. Indeed, a high number of anomalies detected in a neighborhood area may provide enough justification to a more cautious threat assessment, possibly leading to an increased number of Suspect vessel labellings.

Problems may share some attributes and sources (e.g., both anomaly detection and route extraction use the location attribute) but are not necessarily gathered at the same instant in time: Route extraction uses historical data, while maritime anomaly detection uses current data.

The problem outputs can be either used as contextual information (e.g., the routes) or as a problem attribute for the next problem (e.g., anomalies). Also, the same contextual information can be used as an input to different problems, such as the meteorological conditions.

Just as context is defined relatively to the problem at hand, the measurement and decision spaces are also relatively defined. Indeed, the distinction between these two spaces defines the focus of interest (driven by the user's needs) as the decision space. However, a sequential problem would use this decision space as part of its measurement space. For instance, the different types of anomalies detected within Problem B are inputs (measurements) for the threat assessment problem, which aims at inferring some possible intent from these.

Each problem-solving situation contains at least an aggregation (or fusion) function, denoted $\oplus$ that allows us to integrate and draw inferences from information provided by different sources, and an information quality assessment function, whose purpose is to detect conflict, assess, and characterize uncertainty about state estimation.

Conclusions and Future Work

The explicit consideration of context is part of the solution to the challenges of the formulation and implementation of naturally embedded problems. It provides greater flexibility and modularity to the design of information systems for more efficient computation. This idea is central to the development of most of the context-based approaches in information systems across different domains such as artificial intelligence, Big Data, information fusion, and natural language processing. It is expected that a formal consideration of context to the problem of Maritime Situation Awareness would improve the algorithmic solutions to its embedded and interrelated problems such as route extraction, maritime anomaly detection, and threat assessment.

In this chapter, we discussed the benefits of context-based reasoning for Maritime Situation Awareness in the design of information fusion systems. We presented three embedded problems of MSA (route extraction, maritime anomaly detection, and threat assessment) and described their own context. We discussed some benefit of formally introducing the notation of context in the design of MSA support systems within an information fusion perspective, when the variety of sources of information and the associated lack of information veracity is a key challenge. We discussed the system's dimensions influenced by context, and highlighted that context is a relatively defined notion compared to problem-solving situation. While this work is primarily focused on Maritime Situation Awareness where information processing is the core challenge, it can be extended to the information-gathering task, and therefore toward other application domains in which, for instance, the motion pattern of mobile vehicles is driven by an optimized performance of detection or classification (i.e., optimized-source quality).

References

Bouramoul, A., M.-K. Kholladi, and B.-L. Doan. Using context to improve the evaluation of information retrieval systems. *International Journal of Database Management Systems (IJDMS)*, 3(2): 22–39, 2011.

Brezillon, P. Context in problem solving: A survey. *The Knowledge Engineering Review*, 14(1): 1–34, 1999.

Brezillon, P. Making context explicit in communicating objects. In C. Kintzig, G. Poulain, G. Privat, and P.-N. Favennec (Eds.), *Communicating with Smart Objects*, Chapter 21, pp. 273–284. London, UK: Kogan Page Science, 2003.

Bryant, D. Combat identification (15au): Project summary and close out. Technical report TR 2009-128, Defense Research & Development Canada, Toronto, Canada, July 2009.

Department of Defense (DoD). Joint warfighting science and technology plan. Technical report, Deputy Under Secretary of Defense (Science and Technology), February 2000. www.wslfweb.org/docs/dstp2000/JWSTPPDF/00-title.pdf.

Dey, A. K. Understanding and using context. *Personal and Ubiquitous Computing*, 5(1): 4–7, 2001.

Jenkins, M. P., G. A. Gross, A. M. Bisantz, and R. Nagi. Towards context aware data fusion: Modeling and integration of situationally qualified human observations to manage uncertainty in a hard + soft fusion process. *Information Fusion*, 21: 130–144, 2005.

Jousselme, A.-L., and G. Pallotta. Dissecting uncertainty-based fusion techniques for maritime anomaly detection. In *Proceedings of the 18th International Conference on Information Fusion*, Washington, DC, July 2015.

Lane, R. O., D. A. Nevell, S. D. Hayward, and T. W. Beaney. Maritime anomaly detection and threat assessment. In *Proceedings of the 13th International Conference on Information Fusion*, Edinburgh, UK, 2010.

Laxhammar, R. Anomaly detection for sea surveillance. In *Proceedings of the International Conference on Information Fusion*, Firenze, Italy, July 2008.

Llinas, J., C. Bowman, G. Rogova, A. Steinberg, E. Waltz, and F. White. Revisiting the JDL data fusion model II. In *Proceedings of the 7th International Conference on Information Fusion*, pp. 1218–1230, Stockholm, Sweden, June–July 2004.

Llinas, J., A.-L. Jousselme, and G. Gross. Context as an uncertain source. In L. Snidaro, J. Garcia, J. Llinas, and E. Blasch (Eds.), *Context Enhanced Information Fusion: Boosting Real World Performance with Domain Knowledge,* Advances in Computer Vision and Pattern Recognition. Cham, Switzerland:. Springer, 2016.

Pallotta, G., and A.-L. Jousselme. Data-driven detection and context-based classification of maritime anomalies. In *Proceedings of the 18th International Conference on Information Fusion*, Washington, DC, July 2015.

Pallotta, G., M. Vespe, and K. Bryan. Vessel pattern knowledge discovery from AIS data: A framework for anomaly detection and route prediction. *Entropy*, 5(6): 2218–2245, 2013.

Razzaque, M. A., S. Dobson, and P. Nixon. Classification and modeling of the quality of contextual information. In *Proceedings of the IJCAI 2005 Workshop on AI and Autonomic Communications*, Edinburgh, Scotland EJ, 30 July–5 August, 2006.

Snidaro, L., J. Garcia, and J. Llinas. Context-based information fusion: A survey and discussion. *Information Fusion*, 25: 16–31, 2015.

STANAG 1241. *NATO Standard Identity Description Structure for Tactical Use: MAROPS* (5th ed.), April 2005. NATO unclassified.

STANAG 4162. *Identification Data Combining Process* (2nd ed.), December 2001. NATO unclassified.

Steinberg, A., and C. Bowman. Adaptive context exploitation. In J. Llinas, B. D. Broome, and D. L. Hall (Eds.), In *Proceedings of SPIE, Next-Generation Analyst*, Vol. 8758, 2013. Baltimore, MD: SPIE Defense, Security, and Sensing.

Winograd, T. Architectures for context. *Human-Computer Interaction*, 16(2): 401–419, 2001.

Chapter 4

Harnessing Single Board Computers for Military Data Analytics

Suzanne J. Matthews

Contents

Introduction

Recent advances in computer architecture and processor design have given rise to single board computers (or SBCs), where the entirety of the computer is printed on a single circuit board. As the ecosystem of SBCs continues to evolve in compute capabilities and efficiency, the devices become increasingly attractive for military applications, especially for localized data pre-processing and analysis.

Typically, such analysis is conducted by systems such as laptops, portable desktop computers, or via satellite communication with remote high-performance computing (HPC) clusters. In a battlefield environment, the power consumption and cooling requirements of larger systems can be imposing, especially in harsh climates. Dust, high temperatures, and fluctuations in power are all confounding elements. Furthermore, the latency and security requirements of satellite communications can delay the analysis of the information needed for effective command, control, and intelligence (C^2I).

SBCs offer several advantages in the military domain. First, their small form factor and relative inexpensiveness enable high versatility. Second, flash storage enables fast access to data without the latency or power consumption of spinning-disk storage (Cox et al., 2013). Last, their system-on-a-chip (SoC) processors enable data storage capacities and processing capabilities that far outstrip microcontrollers and field programmable gate arrays (FPGAs). SBCs are easily reprogrammable, and have a wide array of ports that enable them to be used as standalone computers or mounted on other devices for a variety of applications.

We predict SBCs will play a vital role in future military operations. The expected growth of battlefield data requires new technology that can efficiently summarize data collected from sensors and other devices into a format that is readable for human operators involved in C^2I activities. We predict the power efficient yet inexpensive SBC will play a critical role in future missions as a "middle man," operating in a hierarchy of devices that work in tandem with microcontrollers and sensors to locally analyze data, consequently expediting the time required for C^2I capabilities.

In this chapter, we present examples of efforts that utilize SBCs for a variety of data analytics activities that have direct application to the military, and we discuss challenges and opportunities for the future use of SBCs. This chapter is not meant to be an exhaustive survey. Instead, we seek to highlight certain key classes of applications for SBCs in the military domain. Notably, we do not cover the use of SBCs for breaking into networks or covertly collecting information (packet sniffing, port scanning, skimming, spoofing, siphoning data, etc.). Instead, our goal is to discuss applications in which SBCs can expedite the analysis of information on-site.

Single Board Computers, Field Programmable Gate Arrays, and Microcontrollers

Before we continue, it is necessary to discuss how SBCs differ from other categories of low-energy chipsets such as microcontrollers and FPGAs. A microcontroller contains only a subset of the functionality of an SBC and is designed to run a single program upon booting. Microcontrollers do not have operating systems and are extremely resource constrained. For example, the Arduino 101 has 24 KB of memory and 196 KB of Flash storage. The Arduino costs approximately $30.00 and has a processor speed of 32 MHz. In contrast, modern SBCs support up to 8 GB of memory, and support high-capacity microSD flash storage.

Field programmable gate arrays (FPGAs) offer a greater level of flexibility than microcontrollers through their circuit reprogrammability but require knowledge of a hardware descriptor language such as VHDL. Programming FPGAs also requires a significant learning curve, even for experienced programmers. In contrast, the Arduino's programming language is very C-like and more accessible for novice programmers. SBCs feature the greatest level of language flexibility of all, with many SBCs programmable in C/C++, Python, and other common programming languages.

Unlike microcontrollers and FPGAs, SBCs are fully functioning computers. They feature operating systems, random access memory (RAM), and power-efficient system-on-a-chip (SoC) processors that are commonly found in smartphones. While microcontrollers and FPGAs can be used in some contexts, we argue that their diminished on-board resources limit their usefulness for many data processing applications required for C^2I.

While smartphones and other mobile devices have similar processors to SBCs, their limited number of ports and lack of general purpose input/output (GPIO) interfaces makes them difficult to integrate into systems that combine hardware and software, such as unmanned vehicles. Common SBCs typically have a wide array of ports. For example, the Raspberry Pi 3 has 4 USB ports, an Ethernet port, a full-sized HDMI port, a micro-USB power connector and a 4-pole A/V header. In contrast, most smartphones only have a single port for charging.

The rest of the chapter discusses common SBCs and how researchers have begun to explore their use for data summarization and analysis applications relevant to the military.

A Snapshot of the Current Ecosystem of SBCs

Table 4.1 contains some popular SBCs ordered according to price. Most of the mentioned SBCs were initially released in the last 5 years. We stress that the listed SBCs represent just a fraction of the ecosystem of available devices. The cheapest of the listed SBCs is just $5.00. The most expensive is $192.00.

Perhaps the most popular SBC in use today is the Raspberry Pi, a credit-card sized computer retailing at $35.00. Initially released in 2012 with a 700 MHz processor and 256 MB of RAM, the Raspberry Pi has enjoyed annual memory and CPU upgrades, while maintaining its $35.00 price-point and form factor. Released in 2016, the Raspberry Pi 3 has 1.2 GHz quad-core ARM processor, 1 GB of RAM, integrated wireless and Bluetooth, and removable microSD flash storage with capacities that can exceed 64 GB.

In January 2014, Intel announced the Intel Edison, a "computer on module" (COM), which is a subtype of SBC. Unlike a regular SBC, a COM must be mounted on a baseboard in order to make use of its I/O interface. In November 2015,

Table 4.1 Popular Single Board Computers

Name	*Price*	*Processor Type (Number of Cores)*	*Memory*	*Dimensions (inches)*	*Power (W)*	*Weight (g)*
Raspberry Pi Zero	$5.00	ARM 11 (1 core)	512 MB	1.18 × 2.56	0.7	9
Raspberry Pi 3	$35.00	ARM A53 (4 cores)	1 GB	3.4 × 2.2	5	42
Adapteva Parallella	$99.00	ARM A9 (2 cores) 16 Epiphany cores	1 GB	2.1 × 3.5	5	42.5
NVidia Jetson TK1	$192.00	ARM A15 (4 cores) 192 Cuda cores	2 GB	5 × 5	58	120

the Raspberry Pi Foundation released the Raspberry Pi Zero, a $5.00 SBC designed to compete with the Intel Edison in the internet of things (IoT) and wearables market. In February 2017, the $10.00 Raspberry Pi Zero W was released. It is a variant of the Raspberry Pi Zero, with integrated wireless and Bluetooth. Intel discontinued the Edison project in early 2017 (Intel, 2017).

The low cost and energy consumption of the SBCs mentioned earlier are largely due to their relatively "weak" CPUs. In addition to an ARM CPU, the Parallella and NVidia Jetson SBCs each feature additional chipsets capable of handling more compute-intensive operations. Released to the general public in 2014, the Parallella is a credit-card sized SBC that boasts a 16-core Epiphany coprocessor while maintaining a similar power profile to the Raspberry Pi 3. NVidia released the Jetson TK1 in May 2014, which contains a 192-core Tegra K1 GPU.

Application 1: Expediting Information Flow in Wireless Sensor Networks

A wireless sensor network (WSN) consists of a collection of sensors that gather information about the surrounding environment and work in tandem to communicate that information back to one or more central resources. Prior surveys (Đurišić et al., 2012; Winkler et al., 2008) have extensively discussed the military applications of WSNs. In the battlefield arena, WSNs are often employed for force protection (e.g., perimeter security/infiltration detection) and monitoring militant activity. For example, a large number of wireless sensors dropped in enemy territory can form a WSN that can indicate enemy presence in remote areas (Malladi and Agrawal, 2002). This increases C^2I capabilities while minimizing risk to ground troops.

SBCs can expedite the information flow in a WSN. A natural location for SBCs in a WSN is at the so-called "gateway" nodes, which act as a central point for information gathering and processing. SBCs can also be used in tandem with microcontrollers to create various workflows. Bell (2013) dedicates an entire book discussing how the Raspberry Pi SBC can be used in conjunction with Arduino microcontrollers for WSN applications. For example, wireless sensors can connect to an Arduino, which can in turn store data on a Raspberry Pi running a MySQL database.

Due to their larger data capacities and more powerful processors, SBCs also enable a greater level of at-node data pre-processing, reducing the amount of raw data that must be transferred across the wireless sensor network. Winkler et al. (2008) note that at-node data pre-processing is critical for energy-efficient wireless sensor networks, as the energy required to transfer data often exceeds the amount required to process the data at the source.

Vujović and Maksimović (2014) explore the feasibility of the Raspberry Pi as a sensor node in a wireless sensor network. In their paper, Vujović and Maksimović compare the Raspberry Pi to five commercial wireless sensors and gathered benchmarks on form factor, power consumption, cost, and memory. While the Raspberry Pi is physically larger than the surveyed commercial wireless sensor nodes, it costs approximately 3–8 times less and has 4000 times the memory. However, the authors note that the Raspberry Pi consumes more power than the commercial wireless sensor nodes, are difficult to power via battery, and lack Bluetooth or integrated wireless capability. That said, it is important to note that this paper was published prior to the release of the Raspberry Pi 3 and Raspberry Pi Zero W, both of which have integrated wireless and Bluetooth. We also note that the Intel Edison (with the Arduino breakout board) has integrated Wi-Fi, Bluetooth, and Ethernet connectivity.

We note the power consumption of both the Intel Edison and Raspberry Pi Zero is around one Watt. While microcontrollers are more power efficient, the additional processing power available at-node may reduce the overall power consumption and latency required to communicate data across the network. PorcupineLabs' PiSense (Porcupine Labs, 2015) is a recent mobile WSN effort that uses Raspberry Pis to collect sensor data. Students at Stanford (Hong et al., 2014) also developed EdiSense, an Edison-based WSN data store designed to help prevent data loss in Bluetooth low energy (BLE) networks.

Application 2: Onboard Data Analysis and Summarization for Unmanned Vehicles

An unmanned vehicle is a machine that moves through and responds to its environment in an unsupervised or semi-supervised manner. Examples include unmanned aerial vehicles (UAVs) and autonomous ground vehicles (AGVs). The military applications of such systems are extensive; UAVs and AGVs enable tactical insight into areas that are dangerous to military personnel, facilitating surveillance and scouting missions and the transport of goods or payload without risk to operator. UAVs and AGVs are commonly used for applications such as terrain mapping, transport (Yamauchi, 2004), and surveillance (Samad et al., 2007).

To enable high mobility, the onboard computer used to gather and summarize sensor data must necessarily be lightweight and power efficient. At the same time, nontrivial computational power is needed to process acquired sensor and image data, which are fed continuously to the onboard computer via mounted cameras. When controlled by a remote operator, the captured data is streamed to a remote location via satellite. We note that SBCs with a single ARM SoC may lack the processing power for such applications. However, SBCs with graphics processing units (GPUs) can accomplish the task while maintaining the small form factor and power efficiency needed for integration with UAVs and AGVs.

The Raspberry Pi has been used extensively in hobbyist drone projects. For example, hardware attached on top (HAT) such as the Navio 2 (Emlid, 2016) can be used in conjunction with the Raspberry Pi and open-sourced UAV platforms such as ArduPilot Mega (APM) (ArduPilot Mega Project, 2017) or PixHawk (Meier et al., 2015) to create drones that are capable of operating in various flight modes and transferring video to a remote operator or network. Traditionally, compute-intensive image processing applications such as feature extraction, video summarizing, object detection, and tracking are delegated to a remote compute system with greater processing power.

Researchers have recently started exploring the extent to which SBCs can perform onboard image processing and summarization. Rebouças et al. (2013) explored how well a Raspberry Pi can analyze captured UAV flight image data. Choi et al. (2016) used an onboard Raspberry Pi and PixHawk to capture the location of a stationary target. Da Silva et al. (2015) studied the suitability of the Raspberry Pi and Intel Edison for onboard aerial image processing to the DE2i-150 FPGA development kit. While the DE2i-150 was over 700 ms faster on average, its excess weight (800 g) and higher energy consumption caused the researchers to conclude that the Pi and Edison were better choices for UAV design. In a separate effort, Vega et al. (2015) compared the efficacy of the Raspberry Pi 2 and the quad-core ARM processor of the NVidia Jetson as an onboard computer for summarizing video data prior to transfer to a remote UAV operator. In this application, the goal is for the onboard computer's frame rate to be equivalent to the observed frame rate by the UAV operator. The researchers demonstrated that by switching to the NVidia Jetson from a Raspberry Pi, they were able to achieve acceptable frame rate results.

A key strength of the NVidia Jetson TK1 is the presence of the onboard Tegra GPU, which has 192 Cuda cores. This makes the board a prime candidate for more intensive onboard image processing applications for autonomous vehicles. Meng et al. (2015) used the NVidia Jetson as an onboard computer for quadcopters deploying their SkyStitch software. For this application, the Cuda cores were used for feature extraction and outlier removal of captured image data. The researchers note that delegating feature extraction to the onboard computer enabled SkyStitch to perform very efficiently. Stokke et al. (2015) discuss how video encoding applications can be conducted with a high energy efficiency using NVidia Jetson TK1 and a "race to sleep" approach for controlling CPU utilization.

The NVidia Jetson has also been used for computer vision applications for autonomous vehicles. Two recent efforts discuss how the NVidia Jetson can be used for real-time lane detection applications. Lee and Kim (2016) tested a custom lane-detection approach on the NVidia Jetson TK1. When benchmarking their results with a CalTech public dataset, they were able to achieve up to 96% accuracy at a frame rate of 44 FPS. Kim et al. (2016) demonstrated how Hough space image transformation can be used for lane detection using the Nvidia Jetson TK1. Sense and Avoidance algorithms are also a critical part of collision prevention and enemy avoidance in autonomous vehicles. Zsedrovits et al. (2015, 2016) implemented sense and avoidance algorithms on the NVidia Jeston, which yielded targeted results.

Application 3: Portable and Localized Cluster Computing for Battlefield Environments

Cluster computing involves a collection of computers networked together to accomplish a common goal. The classic example is the Beowulf cluster, where the individual computers are composed of identical units of low-cost commodity hardware, commonly referred to as commercial off-the-shelf or "COTS." The assembled clusters are loaded with libraries and programs that enable communication and data sharing between nodes. Modern-day clusters are used to create military data centers and HPC research facilities.

While soldiers deployed downrange do not usually need access to HPC systems, compute-intensive applications can be offloaded to a remote HPC center via satellite. Issues with power, security, latency, and wireless network bandwidth can prevent soldiers from maintaining a connection to a remote system or transferring large amounts of data for analysis, especially in a battlefield environment. Due to their high cost, power consumption, and cooling requirements, it is infeasible to deploy traditional HPC systems downrange.

A collection of SBCs can be configured into a Beowulf cluster to create a portable, power efficient compute cluster that can be deployed in a battlefield arena. We stress that SBC clusters are not designed to compete with traditional HPC systems in terms of raw computing power. However, they can serve as an alternative to multi-core desktop computers downrange. The ARM architecture is also expected to play a heavy role in the design of future HPC architectures. Software designed for ARM-based SBC clusters can arguably be ported to future ARM HPC systems as they become available.

A key advantage of using a cluster of SBCs in a battlefield environment is that it removes single points of failure (Cox et al., 2013), which can be disastrous for military applications. Furthermore, the use of an SBC cluster can expedite the processing and analysis of data on-site, enabling the resulting analysis to be accessed immediately. Furthermore, any amount of local data processing and analysis reduces the latency of transfer over satellite communication channels, expediting the speed at which intelligence can be communicated to command centers.

Several researchers have explored the feasibility of SBC clusters for high-throughput data processing applications, specifically in the context of Hadoop MapReduce and Apache Spark. IridisPi (Cox et al., 2013) studied the efficacy of Hadoop's filesystem on a 64-node Raspberry Pi cluster. They note that while their cluster is capable of computing map and reduce tasks, the Hadoop I/O results are very slow. Anwar et al. (2014) studied the performance of various ARM architectures for various Hadoop applications and noted that benchmarking on the 10-node Raspberry Pi cluster was orders of magnitude slower than other architectures. Kaewkasi and Srisuruk (2014) built an Apache Spark cluster out of 22 CubieBoard SBC nodes, which had more powerful ARM Cortex A8 nodes (compared to the older ARMv6 technology implemented in the original Raspbery Pi). Despite the improved CPU, the researchers still concluded that Hadoop I/O issues reduced the efficacy of using SBC clusters for this purpose. More recently, Kachris et al. (2016) studied the power efficiency of Spark applications on various ARM-based architectures, including the Raspberry Pi 3. While they did not build an SBC cluster, they noted that the power efficiency of the ARM architectures makes them attractive for future exploration with Apache Spark.

SBC clusters have shown considerable promise for compute-intensive applications. The goal for such applications is to distribute the total computations over multiple cores and nodes in an effort to reduce application execution time. Distributed computation is enabled through the message passing interface (MPI), an industry standard that is widely used on HPC systems.

Several researchers have used SBC clusters in conjunction with MPI to speed up various computations. The 64-node cluster designed by Cox et al. (2013) used MPI for compute-intensive jobs. Kiepert (2013) built a 32-node Raspberry Pi cluster and benchmarked the performance of a Monte Carlo estimation of Pi (PMCPI) over multiple nodes, compared to a 64-bit Intel Xeon processor. While execution on a single Raspberry Pi node was significantly slower than the Intel processor, PMCPI's run time on 32 nodes was roughly equivalent to its single-thread execution on the Intel Xeon. Matthews et al. (2016) compared the performance of two SBC clusters (Raspberry Pi 2 and Parallella, respectively) against a high-end laptop on the application of password cracking using John the Ripper and MPI. The authors also theorized how the clusters could be applied to other applications in the cyber domain, such as counter-RPA and intrusion detection.

SBC clusters can also be used to run or simulate web servers. Varghese et al. (2014) proposed Raspberry Pi webservers as a green alternative to traditional enterprise solutions at small to medium-sized institutions. The authors found that their Raspberry Pi cluster was able to serve on average 17–23 times more requests per Watt than traditional servers. For static web content, their cluster was able to service up to 200 requests per second, which is sufficient for small websites.

The authors note that for dynamic content, the cluster was suboptimal, achieving only up to 20 requests per second. Overall, however, the authors argue that an SBC cluster is a green alternative to traditional webservers, especially for smaller institutions.

Tso et al. (2013) created a 56-node Raspberry Pi "mini cloud" data center test-bed with the use of Linux containers. Abrahamsson et al. (2013) created a 300-node Raspberry Pi cluster test-bed for experimenting with cloud services. In a more recent study, Pahl et al. (2016) explored the use of SBC clusters in conjunction with containers to serve as "edge-clouds" between IoT devices and larger data centers. Similar to the use of single SBCs in wireless sensor networks, the authors propose that SBC clusters can be used for intermediate data processing for data gathered from IoT devices, reducing the amount of data required to be transferred to a larger data center. Spillner et al. (2015) implemented a "stealth database" on a cluster of 8 Raspberry Pi nodes. Djanali et al. (2014) explored the use of a 10-node Raspberry Pi cluster as a honeypot server for SQL Injection attacks.

Challenges and Opportunities for SBCs in Future Military Operations

The analysis of large amounts of data will play a critical role in future warfare operations. In their article "CyberWar is Coming!" Arquilla and Ronfeldt paint a new picture of war where "light, highly mobile forces" with decentralized information systems provide commanders with "unparalleled intelligence" (Arquilla and Ronfeldt, 1993). More recently, Kott et al. (2016) discuss *The Internet of Battle Things*, in which large numbers of decentralized "intelligent" devices will be gathering and communicating intelligence at a scale previously unseen in modern warfare. A key challenge the article mentioned is the reduction of the vast amount of information produced by devices in a battlefield arena to manageable levels to a summarized format that is meaningful and readable to human actors.

SBCs can be an inexpensive way to achieve these data summarization goals. With the expected deluge of data, we cannot always rely on satellite communication due to security or latency concerns. Setting up local HPC systems for data processing is not an option, due to the high cost required to build, power, and cool such systems, which becomes almost impossible in environments with high temperatures and scarce water resources. Furthermore, local HPC centers are difficult to move and represent a high-value target for the enemy.

Power efficient, decentralized data analytics increases the security and reduces the latency of acquiring C^2I data. SBCs have a distinct advantage over microcontrollers and FPGAs due to their ease of reprogrammability and increased processing speeds and memory capacities. The lack of moving parts in an SBC makes it more ideal for use in harsh climates than standard computers. It also removes a single point of failure since each SBC can easily be replaced.

In the future, it is possible that each soldier has their own personal SBC and a set of microSD cards as part of their standard equipment. Unlike many microcontrollers and FPGAs, most of the mentioned SBCs do not have integrated flash memory. From a security standpoint, this can be advantageous, especially when the goal is to minimize the enemy's ability to capture mission-critical data. Soldiers can swap microSD cards into the device based on needed applications. In the case that data needs to be quickly removed due to the enemy's approach, microSD cards can be easily removed from the SBC, leaving the device behind. Even if equipment needs to be destroyed, it can be done so at reduced expense, given the relative cheapness of microSD cards and SBCs. This is an example of the "disposable security" discussed by Kott et al. (2016).

In visualizing the 2050 battle arena, Kott et al. (2015) state that intelligent warfare will be prevalent, with compact and mobile variants of current systems such as UAVs and "fire and forget" missiles. We have already discussed extensively in this chapter how SBCs can assist with local data processing needs of UAVs by acting as lightweight, yet powerful onboard computers. The use of SBCs for "fire and forget" missiles has also been previously explored by Ramirez et al. (2015). In their paper, they discuss the use of Raspberry Pis in conjunction with the message passing interface (MPI) to create "smart rounds" that can receive independent configuration instructions from a magazine server. In the design, Raspberry Pis are mounted on mortar rounds and wirelessly receive information from the magazine server. We note the paper was published prior to the release of the Raspberry Pi Zero, which can be used to implement the design at even lower cost.

Real technical challenges do exist that limit SBC's current use in tactical environments. For example, some SBCs such as the Raspberry Pi do not have a built-in real-time clock (RTC) module to facilitate low cost. However, an RTC module can easily be purchased separately and wired to the Pi. Other more expensive SBCs, such as the Intel Edison, already have built-in RTC modules.

Perhaps the greatest technical challenge facing SBCs in the military domain is their current inability to be sustained on battery power for extended periods of time. However, there is significant evidence that this hurdle will soon be overcome. In the short term, USB power packs and lithium-ion polymer batteries are being designed with SBCs in mind. In the long term, we predict that SBCs will be powered by newly-invented solid-state batteries. Led by John Goodenough (the inventor of the lithium-ion battery), the new solid-state batteries are inexpensive, provide three times the energy as their lithium-ion counterparts, and can operate in under sub-zero temperatures (Zaragoza, 2017). It is expected that this new technology will lead to longer-lasting rechargeable batteries for handheld devices. We predict that the use of this technology will be crucial to the success of SBCs for future military data processing applications.

Last, we note that the SBC ecosystem is quickly evolving, partially to respond to the demand of the IoT market. ARM chipsets similar to those found in SBCs continue to have lower fabrication costs, ensuring that more powerful SoC will appear on future SBCs. The increasing ubiquity of chipsets with multiple cores and greater Random Access Memory means that programmers can create parallel applications that decrease application run-time and increase data processing capabilities. For example, NVidia recently released the Jetson TX2, a 128-bit SBC with 256 CUDA cores, a quad-core ARM processor. and 8 GB of RAM (NVidia, 2017). It is advertised as a "supercomputer on module" and an "embedded platform for autonomous everything."

Intel has also been working hard to create a viable SBC with an onboard Intel processor. While Intel released a follow-up to its Edison COM called the Intel Joule (Intel, 2016) in 2016, consisting of (a 64-bit SBC with a quad-core Intel Atom CPU and 3 GB of RAM), the project was discontinued less than a year later. Since then, Intel has begun to promote the UP^2, the most advanced models of which have an Intel Pentium Quad-Core processor, 8 GB of RAM, and retails at $319.00 (UP Squared, 2017). More impressively, the UP^2 can operate in temperatures from 32°F to 140°F, making it suitable for deployment in areas with high temperature. One attraction of an Intel processor on an SBC is portability; data processing and summarizing techniques that are created on an Intel laptop should run "as-is" on an Intel SBC, reducing the time between development and deployment.

Conclusions

Single board computers are energy-efficient platforms that are potentially useful for a wide variety of data analytics operations, especially in the military realm. Portability, security, and the ability to operate in potentially power-unstable environments are extremely important for military data analytics applications occurring downrange. SBCs can be incorporated with traditional hardware to create a heterogeneous ecosystem of devices that are capable of performing data summarization and preliminary analysis at every step. SBCs are also lightweight and have a low power profile, enabling them to be incorporated into unmanned aerial vehicles and autonomous ground vehicles. Further data processing capabilities are possible by networking SBCs into clusters.

We are not arguing that individual SBCs or SBC clusters should compete in the same arena as large data centers or high-performance computing resources. When considering performance per Watt or raw compute numbers, larger traditional systems easily beat individual SBCs or SBC clusters (Cloutier et al., 2014). However, SBCs are portable and consume less total energy than traditional systems.

There is also an argument to be made about energy-proportional computing. Barroso and Hölzle (2007) observe that for Google servers, peak energy efficiency occurs at peak utilization. They argue that system designers should focus on developing machines that consume energy in proportion to the work performed (Barroso and Hölzle, 2007). While the researchers were arguing for a fundamental change in server design, Da Costa (2013) suggests an alternative for achieving greater energy efficiency: combining more powerful servers with low power processors such as the Intel Atom and the Raspberry Pi. His experiments show that the incorporation of SBCs in a data center alongside more powerful Intel i7s results in a more energy-efficient system than a typical center containing more homogeneous architecture.

Last, we strongly believe that we are witnessing only the beginning of an "arms race" in the development of SBCs. The ecosystem is evolving at considerable speed; the specific SBC models discussed in this chapter may rapidly become obsolete in the coming years. However, the trend toward the future is obvious: SBCs have the power to transform the way data is transferred and summarized in a battlefield environment. SBCs can be used to support a strategy of localized data processing, reducing the total latency in a network of communicating battlefield devices and increasing the speed at which data is summarized and analyzed for use by human operators and smart devices. For these reasons, we predict SBCs will play a critical role in future warfare operations.

Disclaimer

The views expressed in this work are solely those of the author, and do not necessarily reflect those of the U.S. Military Academy, the U.S. Army, or the U.S. Department of Defense.

References

Abrahamsson, P., Helmer, S., Phaphoom, N., Nicolodi, L., Preda, N., Miori, L., Angriman, M. et al. (2013). Affordable and energy-efficient cloud computing clusters: The bolzano raspberry pi cloud cluster experiment. *IEEE 5th International Conference on Cloud Computing Technology and Science* (*CloudCom*) (pp. 170–175). Bristol, UK: IEEE.

Anwar, A., Krish, K. R., and Butt, A. R. (2014). On the use of microservers in supporting hadoop applications. *2014 IEEE International Conference on Cluster Computing* (*CLUSTER*) (pp. 66–74). Madrid, Spain: IEEE.

ArduPilot Mega Project. (2017). Retrieved from ArduPilot Mega, the Open Source Autopilot, last accessed December 14, 2016: http://www.ardupilot.co.uk/.

Arquilla, J., and Ronfeldt, D. (1993). Cyberwar is coming! *Comparative Strategy*, 12(2), 141–165.

Barroso, L. A., and Hölzle, U. (2007). The case for energy-proportional computing. *IEEE Computer*, 40, 33–37.

Bell, C. (2013). *Beginning Sensor Networks with Arduino and Raspberry Pi* (*Technology in Action*). New York: Apress.

Choi, H., Geeves, M., Alsalam, B., and Gonzalez, L. F. (2016). Open source computer-vision based guidance system for UAVs on-board decision making. *2016 IEEE Aerospace Conference* (pp. 1–5). IEEE.

Cloutier, M. F., Paradis, C., and Weaver, V. M. (2014). Design and analysis of a 32-bit embedded high-performance cluster optimized for energy and performance. *Hardware-Software Co-Design for High Performance Computing* (*Co-HPC*) (pp. 1–8). IEEE.

Cox, S. J., Cox, J. T., Boardman, R. P., Johnston, S. J., Scott, M., and O'Brien, N. S. (2013). Iridis-pi: A low-cost, compact demonstration cluster. *Cluster Computing*, 17(2), 349–358.

Da Costa, G. (2013). Heterogeneity: The key to achieve power-proportional computing. *13th IEEE/ACM International Symposium on Cluster, Cloud, and Grid Computing*, Delft, the Netherlands (pp. 656–662).

da Silva, J. F., Brito, A. V., and Nogueira de Moura, H. (2015). An embedded system for aerial image processing from unmanned aerial vehicles. *2015 Brazilian Symposium on Computing Systems Engineering* (*SBESC*) (pp. 154–157). IEEE.

Djanali, S., Arunanto, F. X., Pratomo, B. A., Studiawan, H., and Nugraha, S. G. (2014). SQL injection detection and prevention system with raspberry Pi honeypot cluster for trapping attacker. *International Symposium on Technology Management and Emerging Technologies* (*ISTMET*) (pp. 163–166). Bandung, Indonesia: IEEE.

Đurišić, M. P., Tafa, Z., Dimić, G., and Milutinović, V. (2012). A survey of military applications of wireless sensor networks. *2012 Mediterranean Conference on Embedded Computing* (*MECO*) (pp. 196–199). IEEE.

Emlid. (2016). *Navio2: Linux Autopilot on Raspberry Pi.* Retrieved from Emlid, last accessed December 6, 2016: https://emlid.com/.

Hong, J., Raymond, J., and Shackelford, J. (2014). *EdiSense: A Replicated Datastore for IoT Data.* Stanford, CA: Stanford University.

Intel. (2016, August). *Discover the Intel® Joule™ Compute Module.* Retrieved from Intel Software Developer Zone, last accessed March 2017: https://software.intel.com/en-us/iot/hardware/joule.

Intel. (2017, November). *Discontinued Maker & Innovation Products.* Retrieved from Intel Software Developer Zone, last accessed November 2017: https://software.intel.com/en-us/iot/hardware/discontinued.

Kachris, C., Stamelos, I., and Soudris, D. (2016). Performance and energy evaluation of spark applications on low-power SoCs. *International Conference on Embedded Computer Systems: Architectures, Modelling, Simulation* (*SAMOS*). IEEE.

Kaewkasi, C., and Srisuruk, W. (2014). A study of big data processing constraints on a low-power hadoop cluster. *2014 International Computer Science and Engineering Conference* (*ICSEC*) (pp. 267–272). IEEE.

Kiepert, J. (2013). *Creating a Raspberry Pi-based Beowulf Cluster.* Boise, ID: Boise State Univerisity.

Kim, H.-S., Beak, S.-H., and Park, S.-Y. (2016). Parallel Hough space image generation method for real-time lane detection. *17th International Conference on Advanced Concepts for Intelligent Vision Systems* (pp. 81–91). Lecce, Italy: Springer.

Kott, A., Alberts, D. S., and Wang, C. (2015). Will cybersecurity dictate the outcome of future wars? *Computer*, 48(12), 98–101.

Kott, A., Swami, A., and West, B. J. (2016). The internet of battle things. *Computer*, 46(12), 70–75.

Lee, Y., and Kim, H. (2016). Real-time lane detection and departure warning system on embedded platform. *International Conference on Consumer Electronics* (pp. 1–4). Berlin, Germany: IEEE.

Malladi, R., and Agrawal, D. P. (2002). Current and future applications of mobile and wireless networks. *Communications of the ACM*, 45(10), 144–146.

Matthews, S. J., Blaine, R. W., and Brantly, A. (2016). Evaluating single board computer clusters for cyber operations. *The 1st International Conference on Cyber Conflict in the United States* (*CyconUS*). Washington, DC: IEEE.

Meier, L., Honegger, D., and Pollefeys, M. (2015). PX4: A node-based multithreaded open source robotics framework for deeply embedded platforms. *2015 IEEE International Conference on Robotics and Automation* (*ICRA*) (pp. 6235–6240). Seattle, WA: IEEE.

Meng, X., Wang, W., and Leong, B. (2015). SkyStitch: A cooperative multi-UAV-based real-time video surveillance system with stitching. *Proceedings of the 23rd ACM International Conference on Multimedia* (pp. 261–270). Brisbane, Australia: ACM.

NVidia. (2017, March). *The NVidia Jetson: The embedded platform for autonomous everything.* Retrieved from NVidia Embedded Systems, last accessed March 2017: http://www.nvidia.com/object/embedded-systems-dev-kits-modules.html.

Pahl, C., Helmer, S., Miori, L., Sanin, J., and Lee, B. (2016). A container-based edge cloud PaaS architecture based on raspberry Pi clusters. *IEEE International Conference on Future Internet of Things and Cloud Workshops* (*FiCloudW*) (pp. 117–124). Vienna, Austria: IEEE.

Porcupine Labs. (2015). *PiSense.* http://www.porcupinelabs.com/pisense/.

Ramirez, Z. J., Blaine, R. W., and Matthews, S. J. (2015). Augmenting the remotely operated automated mortar system with message passing. *CrossTalk*, pp. 12–15.

Rebouças, R. A., Eller, Q. D., Habermann, M., and Hideiti Shiguemori, E. (2013). Embedded system for visual odometry and localization of moving objects in images acquired by unmanned aerial vehicles. *013 III Brazilian Symposium on Computing Systems Engineering* (pp. 35–40). IEEE.

Samad, T., Bay, J. S., and Godbole, D. (2007). Network-centric systems for military operations in urban terrain: The role of UAVs. *Proceedings of the IEEE*, 95(1), 92–107.

Spillner, J., Beck, M., Schill, A., and Bohnert, T. M. (2015). Stealth databases: Ensuring user-controlled queries in untrusted cloud environments. *2015 IEEE/ACM 8th International Conference on Utility and Cloud Computing* (*UCC*), Limassol, Cyprus (pp. 261–270).

Stokke, K. R., Stensland, H. K., Griwodz, C., and Halvorsen, P. (2015). Energy efficient video encoding using the tegra K1 mobile processor. *Proceedings of the 6th ACM Multimedia Systems Conference* (pp. 81–84). New York: ACM.

Tso, F. P., White, D. R., Jouet, S., Singer, J., and Pezaros, D. P. (2013). The glasgow raspberry pi cloud: A scale model for cloud computing infrastructures. *IEEE 33rd International Conference on Distributed Computing Systems Workshops* (pp. 108–112). Philadelphia, PA: IEEE.

UP Squared, (2017). *It's time to get squared: UP Squared.* Retrieved from Up Board website, last accessed November 2017: http://www.up-board.org/upsquared/.

Varghese, B., Carlsson, N., Jourjon, G., Mahanti, A., and Shenoy, P. (2014). Greening web servers: A case for ultra low-power web servers. *2014 International Green Computing Conference* (*IGCC*) (pp. 1–8). Dallas, TX: IEEE.

Vega, A., Lin, C. C., Swaminathan, K., Buyuktosunoglu, A., Pankanti, S., and Bose, P. (2015). Resilient, UAV-embedded real-time computing. *2015 33rd IEEE International Conference on Computer Design* (*ICCD*) (pp. 736–739). New York: IEEE.

Vujović, V., and Maksimović, M. (2014). Raspberry Pi as a wireless sensor node: Performances and constraints. *37th International Convention on Information and Communication Technology, Electronics and Microelectronics* (*MIPRO*) (pp. 1013–1018). IEEE.

Winkler, M., Tuchs, K.-D., Hughes, K., and Barclay, G. (2008). Theoretical and practical aspects of military wireless sensor networks. *Journal of Telecommunications and Information Technology*, 2, 37–45.

Yamauchi, B. M. (2004). PackBot: A versatile platform for military robotics. In *Defense and Security* (pp. 228–237). International Society for Optics and Photonics. doi:10.1117/12.538328.

Zaragoza, S. (2017). Lithium-ion battery inventor introduces new technology for fast-charging, noncombustible batteries. *UTNews: University of Texas at Austin Press Releases*, p. 1.

Zsedrovits, T., Peter, P., Bauer, P., Pencz, B. J., Hiba, A., Gozse, I., Kisantal M. et al. (2016). Onboard visual sense and avoid system for small aircraft. *IEEE Aerospace and Electronic Systems Magazine*, 31(9), 18–27.

Zsedrovits, T., Zarándy, Á., Pencz, B., Hiba, A., Nameth, M., and Vanek, B. (2015). Distant aircraft detection in sense-and-avoid on kilo-processor architectures. *2015 European Conference on Circuit Theory and Design* (*ECCTD*) (pp. 1–4). Trondheim, Norway: IEEE.

Chapter 5

Data Analytics and Training Effectiveness Evaluation

Michael Smith, Susan Dass, Clarence Dillon, and Rodney Long

Contents

Military operations in the twenty-first century have demonstrated that the Army's ability to learn and adapt are critical to success, particularly in today's irregular operations (see Nagl, 2002; Robb and Fallows, 2007 for an overview of this literature). As an organization, the Army's success ultimately rests with the soldiers that execute the mission, and depends on the training and education they receive. The growing pace of social and technological change makes it more imperative to apply the best methods and techniques available to help prepare our soldiers to fight and win the nation's wars. As presented in Peter Senge's *The Fifth Discipline* (2006), "The only sustainable competitive advantage is an organization's ability to learn faster than the competition." Yet, the training burden on soldiers is already overwhelming high, with at least 297 days of training required for every 256 days available for training (116% each year) in 2002 and, by some accounts, more 10 years later (Wong and Gerras, 2015). This chapter describes data analytics methods that are available for the Army and its educational institutions to adapt for best learning practices.

Introduction

In the military training and education communities, current data analytics methods have begun to make an impact on the way that courses are designed, run, and evaluated. In an article for the inaugural issue of the Journal of Military Learning, data analytics represent (or are components of) four of the five elements that comprise the authors' vision for the future of military learning (Schatz et al., 2017). While the military has begun to make strides in this arena, insufficient progress has been made toward the design of a structured method to categorize and implement data measurements in the Army. This chapter describes ongoing work with the United States Army Research Laboratory (ARL) to apply data analytics techniques to the design of courses, evaluation of individual and group performances, and the ability to tailor the learning experience to achieve optimal learning outcomes in a minimum amount of time.

Big Data—the contemporary use of parallel processing to derive value from large-scale, heterogeneous datasets—has begun a transformational shift across society. It has already changed the way that business operates, how academia evaluates performance, and promises to reshape society at large. Exponential increases in computer-processing power and data availability continue to drive the creation of qualitatively new analytic methods, tools, and techniques that have transformative implications for learning research and practices. Data analytics have developed to a point where, for example, computers can teach themselves that cats are important in online videos (Markoff, 2012), and cars have been programmed to self-drive (Greenough, 2015). Moreover, the tools and practices that underlie such innovations are freely available as open-source software and are accessible through free educational resources. Within the past 12 years, the field of data analytics has emerged as a synthesis of computer science and statistics, now both necessary for dealing with complex, data-intensive problems. In this chapter, we explore the impact data analytics can have on course design, evaluation of learner performance (for both individuals and groups), and the ability to refine the learning experience. By providing these tools to the Army's instructional designers, they can optimize learning outcomes throughout the Army education and training community. We describe our methodology for research and evaluation, the fields of learning analytics and educational data mining (LA/EDM), data analytics methods and techniques relevant to learning systems, and a framework for applying these methods and techniques via an illustrative use case. The Army's current computer-based and distributed learning resources (United States Army Program Executive Office for Enterprise Information Systems, 2017; United States Army Training Support Center, 2018) and future training delivery platforms (Sottilare, 2017) offer ample data opportunity to leverage LA/EDM across the Army. Ultimately, our goal is to provide a vision for the successful application of data analysis techniques throughout the Army learning community.

Approach

Before discussing how these new techniques can be applied to Army learning, we review current research in the fields of learning analytics (LA) and educational data mining (EDM). We rely on Siemens's review of LA (Siemens, 2012) and Springer's compilations on LA/EDM together (Larusson and White, 2014; Papamitsiou and Economides, 2014; Peña-Ayala, 2013; Siemens and Baker, 2012). The bulk of this chapter categorizes common analytic techniques and evaluates the potential of each category to improve Army learning practices and initiatives. We refer to several key textbooks used to teach data science, such as those by Provost and Fawcett (2013), Chambers and Dinsmore (2015), Loukides (2011), and Schutt and O'Neil (2013). However, many innovations and emerging methods for data science are developed by practitioners and industry researchers rather than peer-reviewed academic publications. We also review recent approaches shared on data analytics websites while writing this chapter, such as O'Reilly Media (2017), KD Nuggets (2017), Data Science Central (2017), and MIT Technology Review (2017). We derive seven categories of methods and develop working definitions for each:

- *Predictive modeling*: Developing of a statistical relationship between input and output variables used to predict a future outcome.
- *Machine learning*: Training computers as intelligent agents to extract models from data without explicit instruction; used to augment other methods. Includes "deep learning."
- *Similarity grouping*: Grouping and retrieving similar data objects based on the measurement of their statistical distance.
- *Recommender systems*: Determining the affinity between a user and a piece of content with the goal of pointing the user to their next desired content.

- *Social network analysis*: Evaluating the relationships and structures of a network to identify central and isolated individuals (and groups) to make predictions.
- *Natural language processing*: Tagging, evaluating similarity, deriving meaning, and processing text data, often as an input to other methods and techniques.
- *Big Data* (and MapReduce): Programming standards and tools popularized by Google—widely used as a method to process large-scale, heterogeneous datasets.

Learning Analytics and Educational Data Mining

Data mining (including machine learning) began in the business industry to support data-driven decision making that relied on sophisticated algorithms at the intersection of computer science and statistics (Papamitsiou and Economides, 2014; Scheffel et al., 2014). EDM then evolved from general data mining to explore methods as applied specifically to educational data to meet educational goals (Chatti and Dyckhoff, 2012), and can be conceptualized as the intersection of statistics, computer science, and education (Romero and Ventura, 2013). EDM is focused on using large datasets to inform education at the institutional level and above, using tailored data mining analytics that initially rely on automated discovery to find patterns. In contrast, LA applies statistical patterns (whether the pattern has been discovered with EDM or with human interpretation) to education goals and supports learners within an institution. The primary connection between EDM and LA is the use of models, whether predictive, descriptive, or prescriptive, as shown in Figure 5.1.

LA is a fast-growing, multidisciplinary field that supports educational research and educational understandings (Dawson et al., 2014; Ferguson, 2012; Scheffel et al., 2014). LA draws from the learning sciences, data processing, web analytics, psychology, philosophy, sociology, linguistics, information visualization, adaptive and

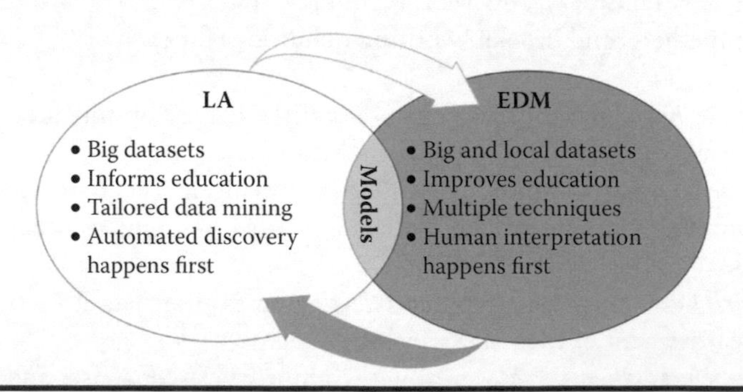

Figure 5.1 The connection between EDM and LA.

adoptive systems, and recommender systems (Chatti and Dyckhoff, 2012; Dawson et al., 2014; Ferguson, 2012; Gasevic et al., 2015; Scheffel et al., 2014). The field of LA continues to evolve from a focus on technological perspectives to a focus on educational perspectives with an anticipation that this evolution will further extend into other disciplines as appropriate (Dawson et al., 2014; Ferguson, 2012).

Together, "LA and EDM constitute an ecosystem of methods and techniques (in general procedures) that successively gather, process, report and act on machine-readable data on an ongoing basis in order to advance the educational environment and reflect on learning processes" (Papamitsiou and Economides, 2014, p. 49). Some researchers perceive this multidisciplinary ecosystem as a positive in that each community adheres to different standards and values in determining what is important to the community and what constitutes good research (Siemens and Baker, 2012). LA and EDM are not limited to computer-based educational environments (the intersection of computer science and education) (Romero and Ventura, 2013), but can be applied together whenever there is quantitative information about Army learners: as individuals, and for squads, schoolhouses, and the department-level of the Army.

LA/EDM stakeholders cover a broad range of roles. Several researchers (Ifenthaler and Widanapathirana, 2014; Romero and Ventura, 2013; Siemens and Baker, 2012) have developed a hierarchical framework to represent group roles. As shown in Figure 5.2, the mega-level analytics represents governance; basically

Type of analytics	Number of students addressed	Hierarchy level and stakeholders
Academic analytics		**Mega-level** National and international Policy and decision makers
		**Macro-level** Regional and state Administrators and policy and decision makers
		**Meso-level** Institutional Administrators, funders, and marketers
Learning analytics		**Micro-level** Individual, small group Learners, faculty, instructional designers, and researchers

Figure 5.2 The hierarchical framework to represent group roles. (From Ifenthaler, D., and Widanapathirana, C., *Technol. Knowl. Learn.*, 19, 221–240, 2014; Shum, S. B., Learning analytics, Technical report, UNESCO Institute for Information Technologies in Education, Moscow, Russia, 2012; Siemens, S. et al., Open learning analytics: An integrated and modularized platform, in *Knowledge Creation Diffusion Utilization*, 1–20, 2011.)

reviewing data at the national and international level by policy and decision-makers. The macro-level analytics explores student data at the level of the military service and across the Department of Defense to evaluate and benchmark against educational goals. The meso-level analytics explores student data at the institution level; at a service academy or a training center, for example. Institutions may be interested in predicting accession, improving student retention, improving learner success, and identifying at-risk students (Campbell et al., 2007). Last, the micro-level addresses individual students and small groups. Learning analytics focused on the micro-level benefits students and faculty, while academic analytics supports the meso- and macro-levels, benefiting administrators, funding agencies, policymakers, and government overseers (Siemens et al., 2011; Shum, 2012).

Related Work in Data Analytics

Data analytics is a scientific process of transforming data into insight for making better decisions (INFORMS, 2017). It is used in industry to improve organizational decision making and in the sciences to verify or disprove existing models or theories. Modern analytic techniques draw from multiple disciplines such as statistics, artificial intelligence, software engineering, and others to solve data-intensive problems and generate novel insights and products. In this section, we describe our seven categories of analytic methods along with potential learning applications for each for LA/EDM.

Predictive Modeling

Predictive modeling in data analytics refers to the use of statistical techniques that allow analysts to leverage the relationship of input and output variables to predict a future outcome. These relationships allow analysts to predict, classify, and act on an outcome prior to the occurrence of an event. Beyond simply outcome prediction, predictive modeling allows analysts to identify the components that influence a specific outcome, enabling targeted interventions to improve performance. When applied to the educational setting, these techniques prove particularly powerful in identifying students at risk of failing or in need of intervention. It can also be used to predict the cost of future training based on the proposed elements.

The current literature on predictive modeling in the academic setting describes its use in determining student success. Kongsakun (2013) describes the use of linear regression models together with clustering techniques to produce a model called e-Grade that predicts a student's likely final course score before and after midterm exams. This model uses predictor variables such as attendance at the first class and average prior grades to assess a student's possible course grade. Students with low predicted course grades are then targeted for intervention by tutors and professors.

Smith et al. (2012) describe a similar process in predicting student success in online community colleges in the United States. Similarly, Hung and Zhang (2006) describe the use of regression models to predict academic performance of students based on online learning logs. These logs contained a range of data, such as reading or posting in online discussion boards, as input variables to predict the success of undergraduate students. Likewise, predictive classification techniques were applied by Lam-On (Lam-On and Boongoen, 2014) to predict student drop-out rates.

Alternatively, Hutzler et al. (2014) describe the use of decision trees to predict and classify the level of difficulty of test questions provided for reading comprehension based on a set of training data in which the test question difficulty for each question has been previously assessed. In addition, the decision-tree model used nine predictor variables converted into a numeric output ranging from question type (e.g., 1 = multiple choice, 2 = open-ended) to presentation of information (e.g., 1 = textual, 2 = graphic). Using the decision-tree model, the learning organization was able to predict the difficulty of new questions and create well-balanced and aligned exams. These types of models are often combined through an automated analytics workflow (a series of procedures that can be efficiently executed in unison), but they can be employed individually as well.

Machine Learning

Machine learning (ML) is a field of practice adapted from the artificial intelligence discipline that focuses on training computers as intelligent agents to extract models from data without explicit programming. While considerable overlap exists across each of the following techniques, the primary focus in ML is on the improvement of a machine's ability to model data based on experience, as opposed to a more generalized development of models where the primary agent is an analyst. In most cases of the techniques we describe here, ML can be used to enhance or improve the process; in other cases, ML can speed up the analytic process. ML should be seen as a tool for tackling problems that would be intractable or inefficient without computer assistance. The greater the complexity of the problem set, the more appropriate to focus on applying machine learning to assist in dimensionality reduction or identifying patterns that may not be readily apparent (Marsland, 2009).

As a supporting method, machine learning has been applied to a wide array of learning applications in conjunction with many of the techniques outlined earlier. One example of this is a popular application known as Support Vector Machines (SVMs) (Steinwart and Christmann, 2008). This machine learning classifier works similarly to other predictive modeling methods by making a binary class prediction for a given subject using non linear functions to fit the model across many dimensions. As a consequence, the predictions are highly accurate but extremely difficult to interpret—a common issue among machine learning applications. In support of ARL, Charles River Analytics has applied SVM to conduct automatic classification of training documentation according to their developmental categories (e.g., *Bloom's Taxonomy*) at a level of accuracy that rivals human annotation. Other applications include those displayed in the KDD Cup 2010, a data mining competition that ranked participants based on the accuracy of a successful prediction model for students in a basic math course (Algebra I). Many of the top-scoring models utilized ensemble methods, techniques that weight the predictions of numerous weaker predictive models to generate an overall score that is more accurate than the individual predictions.

Machine learning techniques can be further divided into supervised and unsupervised processes. Supervised techniques require both input data and some target representing the solution space ("Given some experience about data that does and does not fit this category, does this new data fit the category?"). Unsupervised techniques look for a structure in the relationships within a dataset in order to uncover patterns that might not be expected ("Given some data, how could it be categorized?").

Similarity Grouping

Similarity grouping in data analytics consists of using automated methods to identify and quantify meaningful segments in data based on statistical attributes. Clustering, one of the most popular techniques, uses a variety of attributes to cluster information into meaningful categories based on groupings of high similarity. In clustering techniques, data points grouped together not only share a high similarity within their assigned cluster grouping, but also display a high degree of dissimilarity from other clustered groupings. This is an unsupervised technique: categories are not predefined but are identified by clusters within a dataset. This approach allows analysts to draw new insights from complicated datasets and provide context through the grouping of similar students, courses, and course materials.

Similarity grouping has been used in a variety of contexts when applied to learning analytics and educational data mining. Govindarajan et al. (2013) describe the use of continuous clustering techniques to organize students into similar groupings for the purpose of providing targeted learning objects. Using this clustering technique, students that share similar knowledge gaps based on previous course work or based on in-module knowledge checks can be grouped together to form follow-on classes, form break-out groups with the instructor, or take additional modules in asynchronous training. Likewise, Kumar and Ramaswami (2010) highlight the valuable insights a training institution can gain into the makeup of their student population by clustering by variables outside of course performance including social, cultural, and economic measures. Finally, Abukhousa and Atif (2015) describe the use of cluster models to group similar students into communities of practice that allow like-minded students to share their own industry knowledge outside of the formal course setting.

Valsamidis et al. (2012) and Cui et al. (2005) describe the use of cluster models to group similar courses and course materials with the purpose of gaining a deeper understanding of content or characteristic overlap and for use in combination of additional analytical techniques such as recommender engines. Valsamidis et al.

demonstrate that by using metadata and weblog data of course concepts, training organizations can group similar courses outside of intuitive categories for targeted action. For example, apart from simply subject areas, courses can be grouped into high or low activity groups, allowing organizations to prioritize courses in need of revision or understand the characteristics that define a popular curriculum. Similarly, Cui et al. provide an example of using cluster analysis to organize large libraries of supporting documents to provide students an easier entry into choosing supporting course materials.

Recommender Engines

Recommender engines in data analytics have become a key tool in bringing users the information they seek, perhaps before they begin searching for it. Applied to a learning environment, recommender engines support students in finding the appropriate course, instructors in choosing the most relevant material, and course developers in choosing the best-aligned learning object within large educational databases with relative ease and efficiency. This ability to find the right educational resource at the right time is especially appropriate in today's changing learning environment, as web-based training content databases grow and learners find they are unable to devise the search terms necessary to sufficiently filter results through a simple query. Moreover, "users do not have a precise enough understanding of what they want to formulate specific queries" (Kumar et al., 2007) and may become overwhelmed at the "hit-shock" of receiving an overwhelming number of search results. Recommender engines can quantify the affinity between a user and a content point with the goal of pointing the user to the desired content (Melville and Sindhwani, 2010). They generally apply two primary analytical techniques, content-based filtering and collaborative filtering, though modern approaches typically combine both techniques for large-scale applications. As education increasingly moves to a learner-centric approach (Imran et al., 2015) where the educational path is driven by the learner, recommender engines can assist students, instructors, and teaching organizations in connecting the right content to the right user at the right time.

Kumar et al. (2007) describes the use of recommendation techniques to assist learners in choosing specific learning objects within large databases. Through this technique, the associated course metadata is used to recommend courses of aligned keywords based on a community-filtering model. Alternatively, El-Bishouty et al. (2014) present the use of recommendation techniques to assist learners in choosing learning content based on the student's individual learning style as captured within a survey. Imran et al. (2015) describe the use of recommendation systems in assisting students in self-directed learning programs to select the most relevant tasks, as opposed to content or materials, for their desired training outcome. Within this system, Imran et al. argue for the use of a rule-based recommendation system based on past performance of similar users and course difficulty levels.

Social Network Analysis

Social Network Analysis (SNA) is a principal method for any quantitative social research because it focuses on relationships—the essence of society. SNA is a formal depiction of a social environment expressed as patterns or regularities in relationships among interacting units (Wasserman and Faust, 2009). There exists a vast literature for SNA from both a mathematical-theoretic perspective—graph theory and statistics—as well as a social, application-level perspective. Many of the techniques developed to analyze social networks can be applied to the analysis of other networks—almost anything that can be described by relationships. Most of the common SNA measures derive from simple calculations or statistics about the ratios of nodes to edges for subcomponents of a network.

One example of how SNA is being applied to learning analytics is the Social Networks Adapting Pedagogical Practice (SNAPP) program lead by Shane Dawson from the University of Wollongong (Australia), which uses student social networks to inform instructors about challenges and opportunities among their students in real time (Dawson, 2010; Dawson et al., 2010). The research for this program takes advantage of Dawson's long treatment of the subject of computational learning analytics and SNA, in particular. Other research in this area has included: learner isolation (McDonald and Noakes, 2005); how networks impact creativity (Dawson et al., 2010); and how SNA supports the instructor's view of the "big picture" of large classrooms (including recommendations on content scaffolding as class size grows [Brooks et al., 2007]).

Natural Language Processing and Text Analysis

Natural language processing (NLP) and text analysis (TA) are analytic methods used to extract information from (typically) unstructured texts. Though related, these analytic methods are distinct from text mining (TM), which is more about extracting specific information (data) from a text. Both NLP and TA are often used in conjunction with machine learning, and popular algorithms can be also be found in texts on machine learning and artificial intelligence.

Though there is a distinction between the purpose that NLP and TA each serve, these fields share some of the same evolutionary roots. Both can also be applied to Big Data, with appropriate modifications. The TA techniques focus on text as raw data (word counts, semiotics, roots, etc.) while NLP attempts to extract meaning from sentences or whole documents (semantics, rhetorics, hermeneutics) by leveraging the structure of the text's language. Many techniques have been codified in software libraries and are commonly used as subprocesses together with other machine learning algorithms.

Computational analysis techniques that evolved from this approach can be applied directly to learning analytics. Some practically achievable objectives are: to evaluate media resources that represent or partially represent learning objects and catalog it into machine-readable data; real-time (or near real-time) evaluation of students' (virtual) classroom discussions, questions, assignments, and so on, for sentiment, structure, content, and complexity; and to support careful reading of students' written products for in-depth evaluation to detect plagiarism, identify key concepts, and so forth.

Techniques such as these have already being applied in learning analytics research for several years. In 2001, Wang et al. published a paper on successful discourse analysis of online classroom chats to predict student performance. Wang and his cohort had to code their data by hand. As computer science advances, the possibility of automated coding becomes more feasible (Wang et al., 2001). Just 7 years after Wang's paper, Rosé et al. (2008) published a study comparing computer-based NLP to hand coding, showing between 42% and 97% accuracy over various measures of similarity, using a variety of NLP methods. Data analysis tools and technology have continued to improve since 2008 and now offer a wider selection of methods than Rosé et al. had available.

State-of-the-art TA and NLP methods have improved to the point that (in some subjects) computer-automated coding is no more prone to error than human coding (Leetaru and Schrodt, 2013). However, some features of natural language, such as sentiment and humor, are still difficult to detect with common algorithms and simple machines (Davidov et al., 2010). Similarly, while it is currently possible to train a computer to recognize images in pictures or video, they do not recognize the meanings in those images. Thus, despite advancements, many media types still need to be hand coded.

Big Data Tools and MapReduce

We have already alluded that data analytics with Big Data is a special case, though it is inextricably linked to the methods already discussed. Big Data refers to datasets that are too large or complex to fit in the memory of typical computer workstations and laptops or through traditional relational database methods. The most commonly applied solution, invented by Jeffrey Dean and Sanjay Ghemawat (2004) at Google, is to break the Big Data into small chunks for processing with many computers in parallel—the "map" function—then, "reduce" the volume, velocity, and variety of data. It is often collected from people's everyday activities and computer interactions, and referred to as "data exhaust." The scale of Big Data—the volume and velocity—comes primarily from automatic collection of online events or machine sensor data. The variety of Big Data comes largely from aggregating multiple data streams that were not originally intended to be used together.

The decision to apply Big Data techniques to learning analytics requires an upfront decision to capture and store the interactions students have as part of their computer-based education and training, including any interactions and communications wherever possible. Big Data collection and storage requires unique (though ubiquitous) computing resources. Collection mechanisms need to be written into the learning platform software. Storage and computation typically happen in the cloud to provide data surety and rapid analysis. A Big Data collection strategy might include traditional events like the webpage visits, media downloads, forum post metadata, and so on. The benefit of Big Data for instructors, institutions, and policymakers are manifold. It provides an ability to discover ways that students approach learning and a passive feedback mechanism to judge changes in policies at all levels.

Learning Analytics System for the Army

This section describes an application model for practitioners and organizations to evaluate use cases in the context of a learning system, shown in Figure 5.3. We considered applications in the context of formal, online learning to provide the most data-rich context for evaluating data analysis methods, though this is not intended to preclude application to informal or offline settings. Our review of data analytics applications uncovered three categories of popular usage: content development, real-time support, and post-evaluation. In a content development application, the instructor or instructional designer is assumed to search within an existing repository for content to create a new online learning experience aligned with the desired learning outcomes best suited for the target audience. The desired content may be for an entire course, for a topic, or simply for an instructional activity. A real-time support system continuously analyzes learner performance based on responses and interactions to support the learner as well as the instructor. Real-time analysis could autogenerate remedial support for an at-risk student, related concepts for an advanced student, or could identify problematic or difficult areas within the course. Post-evaluation cases focus on rolling-up

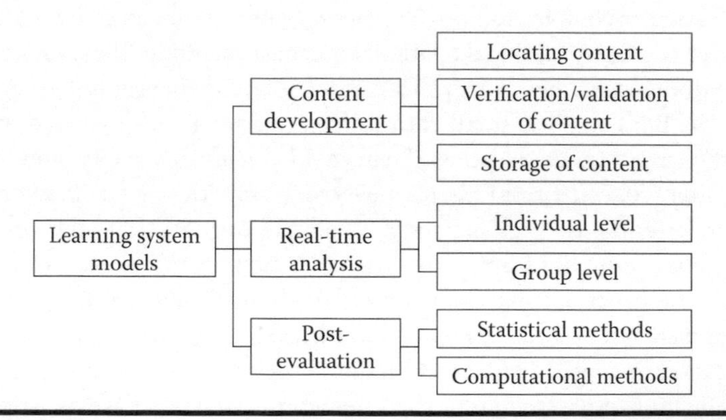

Figure 5.3 An application model for evaluating use cases in the context of learning system modules.

individual and course insights developed during real-time analysis across courses to provide organizational insights. This application applies Big Data principles for integration across institutional data sources to support automated curriculum and schoolhouse evaluation for all levels of the Kirkpatrick/Phillips Evaluation Model (Kirkpatrick, 1998); for example, identifying correlations between student learning and promotion potential using personnel records.

Of the three use categories, real-time support is a particularly interesting use case to explore and gain insight into potential benefits. We imagine a real-time learning analytics system evaluating learner performance along with real-time learner activity. Multiple data-analysis techniques could be used to support real-time analysis as indicated in Table 5.1 and serve multiple purposes. Such analysis could identify learners at risk of poor performance and enable timely

Table 5.1 Data Analytics and Applications

Method	*Real-Time Application*
Recommenders	Identify additional content or resources Locate other learners to connect with
Similarity Groups	Identify learner population segments, composition, and demographics Cluster to identify support for populations and subgroups Explore data to support other data analytic engines
ML	Refine content segmentation and classification Develop learning topics based on current needs
SNA	Assess level of engagement, isolation, and influence Support evaluation of instructor and student performance Monitor peer relationships for requesting help
NLP & TM	Analyze student assignments; support automated feedback Identify plagiarism Evaluate communications for sentiment, e.g., boredom, frustration, etc.
Predictive Models	Assess at-risk, based on assignments, exam scores, exam questions Alter course pace based on learner progress Support intelligent tutoring and instructional intervention
Big Data	Enable comparisons between highly variable data types Support real-time analysis for high-volume, high-velocity data

support and remediation as appropriate. A real-time system might also identify where knowledge already exists and enables advanced learners to skip ahead (reduce the duration of training) or to supplement basic knowledge with advanced concepts (eliminating future training requirements). Real-time analysis across the class could identify problematic or difficult areas within the course so that an instructor can self-correct, as needed. Training modifications might include content reorganization or re-augmentation of content, updating learning objectives, or changing learning activities to improve learner performance.

Real-time analysis is attractive to improve learning effectiveness, but interpretation of the analysis needs requires caution. Consider a case where an algorithm tags a learner as "at-risk" because of insufficient time online reviewing material; adding remedial content is unlikely to resolve the problem. The learner's behavior could stem from lack of motivation, poor time-management skills, sense of isolation, or lack of self-regulation skills as opposed to not understanding the content. A mix of real-time analyses across multiple variables is required, not only to assess and predict learner performance, but also to indicate an appropriate solution.

For example, SNA can be used to assess learner engagement, measured by the volume (count) and frequency (timeline) of contributions to online activities. For one example of how SNA is combined with other data analytics technologies to support online education, consider the Wikispaces website (Tangient LLC, 2017) and their Engagement dashboard element. This gauge shows each classroom participant, their relative connectivity, and their level of activity (forum contributions, Wiki edits, etc.) expressed as volume and frequency. The SNA metric of connectedness informs the viewer whether a student's activity is "just noise" or if fellow students are listening. Exploring these additional activities through sentiment analysis could further determine if the initial posting was perceived as positive or negative.

Real-time analysis could indicate that the course does need additional content to support difficult content. In this case, the instructor may search a learning repository through a dashboard, relying on different criteria to identify and assess the appropriateness of the retrieved content. Appropriateness might be based on learner needs (both relevance and context), based on course content needs (content type, learning difficulty, learning domain, subject domain, and learning level), and/or based on instructional needs (quality, instructional activity, duration, and format). These three themes representing the learner, the content, and the instruction are the important elements in the instructional design of a course (Anderson et al., 2001; Gagne et al., 1988; Merrill, 1994; Ross et al., 2007).

Conclusions

This chapter summarized the fields of learning analytics and educational data mining that are relevant to military learning systems. We developed a framework to evaluate data analytics techniques for military learning systems and explored it in three use cases: content development, real-time analysis, and post-evaluation. Expanding on the real-time analysis use case, we identified and discussed several applicable data analytics techniques in terms of their application to learning analytics. Each of these techniques and applications serve as examples of the potential value of data analytics to support Army learning.

References

Abukhousa, E., and Y. Atif. Big learning data analytics support for engineering career readiness. *Proceedings of 2014 International Conference on Interactive Collaborative Learning, ICL 2014*, pp. 663–668, 2015.

Anderson, L. W., D. Krathwohl, P. W. Airasian, K. A. Cruikshank, R. E. Mayer, P. R. Pintrich, J. Raths, and M. C. Wittrock. *A Taxonomy for Learning, Teaching, and Assessing: A Revision of Bloom's Taxonomy of Educational Objectives Complete Edition.* Pearson, Allyn, & Bacon, New York, 2001.

Brooks, C., W. Liu, C. Hansen, G. McCalla, and J. Greer. Making sense of complex learner data. July:1–42, 2007.

Campbell, J. P., P. B. Deblois, and D. G. Oblinger. Academic analytics: A new tool for a new era. *Educause Review*, 42(4):41–57, 2007.

Chambers, M., and T. W. Dinsmore. *Advanced Analytics Methodologies: Driving Business Value with Analytics*. Pearson Education, Upper Saddle River, NJ, 2015.

Chatti, M. A., and A. L. Dyckhoff. A reference model for learning analytics. *International Journal of Technology Enhanced Learning*, 9:1–22, 2012.

Cui, X., T. E. Potok, and P. Palathingal. Document clustering using particle swarm optimization. *Proceedings 2005 IEEE Swarm Intelligence Symposium, 2005. SIS 2005*, pp. 185–191, 2005.

Data Science Central. http://www.datasciencecentral.com/. Accessed: September 11, 2017.

Davidov, D., O. Tsur, and A. Rappoport. Semi-supervised recognition of sarcastic sentences in Twitter and Amazon. *14th Conference on Computational Natural Language Learning*, July:107–116, 2010.

Dawson, S. "Seeing" the learning community: An exploration of the development of a resource for monitoring online student networking. *British Journal of Educational Technology*, 41(5):736–752, 2010.

Dawson, S., A. Bakharia, and E. Heathcote. SNAPP: Realising the affordances of real-time SNA within networked learning environments. *Proceedings of the 7th International Conference on Networked Learning*, pp. 125–133, 2010.

Dawson, S., D. Gasevic, G. Siemens, and S. Joksimovic. Current state and future trends: A citation network analysis of the learning analytics field. *Proceedings of the Fourth International Conference on Learning Analytics and Knowledge: LAK'14*, pp. 231–240, 2014.

Dean, J., and S. Ghemawat. MapReduce: Simplified data processing on large clusters. *Proceedings of 6th Symposium on Operating Systems Design and Implementation*, pp. 137–149, 2004.

El-Bishouty, M. M., T.-W. Chang, S. Graf, Kinshuk, and N.-S. Chen. Smart e-course recommender based on learning styles. *Journal of Computers in Education*, 1(1):99–111, 2014.

Ferguson, R. Learning analytics: Drivers, developments and challenges. *International Journal of Technology Enhanced Learning*, 4(5/6):304–317, 2012.

Gagne, R. M., L. J. Briggs, and W. W. Wager. *Principles of Instructional Design*, 3rd ed., pp. 177–197. New York: Harcourt Brace Jovanovich, 1988.

Gasevic, D., S. Dawson, and G. Siemens. Let's not forget: Learning analytics are about learning. *TechTrends*, 59(1):64–71, 2015.

Govindarajan, K., T. S. Somasundaram, V. S. Kumar, and Kinshuk. Continuous clustering in big data learning analytics. *Proceedings – 2013 IEEE 5th International Conference on Technology for Education, T4E 2013*, pp. 61–64, 2013.

Greenough, J. The self-driving car report: Forcassts, tech timelines and the benefits and barriers that will impact adoption. *Business Insider*, p. 5, June 2015.

Hung, J.-L., and K. Zhang. Data mining applications to online learning. In T. Reeves and S. Yamashita (Eds.), *Proceedings of E-Learn: World Conference on E-Learning in Corporate, Government, Healthcare, and Higher Education 2006*, pp. 2014–2021, Honolulu, HI, October 2006. Association for the Advancement of Computing in Education (AACE).

Hutzler, D., E. David, M. Avigal, and R. Azoulay. Learning methods for rating the difficulty of reading comprehension questions. In *Proceedings – 2014 IEEE International Conference on Software Science, Technology and Engineering, SWSTE 2014*, pp. 54–62, 2014.

Ifenthaler, D., and C. Widanapathirana. Development and validation of a learning analytics framework: Two case studies using support vector machines. *Technology, Knowledge and Learning*, 19(1–2):221–240, 2014.

Imran, H., M. Belghis-Zadeh, T.-W. Chang, Kinshuk, and S. Graf. Emerging issues in smart learning. In G. Chen, V. Kumar, Kinshuk, R. Huang, and S. C. Kong (Eds.), *Lecture Notes in Educational Technology*. Springer-Verlag, Berlin, Germany, 2015.

INFORMS. Best definition of analytics. https://www.informs.org/About-INFORMS/News-Room/O.R.-and-Analytics-in-the-News/Best-definition-of-analytics, 2017.

KD Nuggets. http://www.kdnuggets.com/. Accessed: September 11, 2017.

Kirkpatrick, D. L. *Another Look at Evaluating Training Programs*. American Society for Training & Development, Alexandria, VA, 1998.

Kumar, V., J. Nesbit, P. Winne, A. Hadwin, D. Jamieson-Noel, and K. Han. *Quality Rating and Recommendation of Learning Objects*, pp. 337–373. Springer, London, UK, 2007.

Lam-On, N., and T. Boongoen. Using cluster ensemble to improve classification of student dropout in Thai university. In *2014 Joint 7th International Conference on Soft Computing and Intelligent Systems, SCIS 2014 and 15th International Symposium on Advanced Intelligent Systems, ISIS 2014*, pp. 452–457, 2014.

Larusson, J. A., and B. White. *Learning Analytics: From Research to Practice*. Springer, New York, 2014.

Leetaru, K., and P. A. Schrodt. GDELT: Global data on events, location and tone, 1979–2012. *Annual Meeting of the International Studies Association*, April:1979–2012, 2013.

Loukides, M. *What Is Data Science?* O'Reilly Media, Sebastopol, CA, 2011.

Markoff, J. How many computer to identify a cat? *New York Times*, p. B1, June 26, 2012.

Marsland, S. *Machine Learning: An Algorithmic Perspective*. Chapman & Hall/CRC, The R Series. CRC Press, Boca Raton, FL, 2009.

McDonald, B., and N. Noakes. Breaking down learner isolation: How social network analysis informs design and facilitation for online learning. In *Proceedings of the American Educational Research Association Conference*, pp. 1–30, 2005.

Melville, P., and V. Sindhwani. *Recommender Systems*. Springer, New York, 2010.

Merrill, M. D. *Instructional Design Theory*. Educational Technology Publications, Englewood Cliffs, NJ, 1994.

MIT Technology Review. https://www.technologyreview.com/. Accessed: September 11, 2017.

Nagl, J. A. *Counterinsurgency Lessons from Malaya and Vietnam: Learning to Eat Soup with a Knife*. Praeger, Westport, CT, 2002.

O'reilly. https://www.oreilly.com/. Accessed: September 11, 2017.

Papamitsiou, Z. K., and A. Economides. Learning analytics and educational data mining in practice: A systematic literature review of empirical evidence. *Journal of Educational Technology & Society*, 17(4):49–64, 2014.

Peña-Ayala, A. *Educational Data Mining: Applications and Trends*. Studies in Computational Intelligence. Springer International, Cham, Switzerland, 2013.

Prakash Kumar, S., and K. S. Ramaswami. Fuzzy k-means cluster validation for institutional quality assessment. In *2010 International Conference on Communication and Computational Intelligence (INCOCCI)*, pp. 628–635. IEEE, 2010.

Provost, F., and T. Fawcett. *Data Science for Business*. O'reilly Media, Sebastopol, CA, 2013.

Robb, J., and J. Fallows. *Brave New War: The Next Stage of Terrorism and the End of Globalization*. John Wiley & Sons, Hoboken, NJ, 2007.

Romero, C., and S. Ventura. Data mining in education. *Wiley Interdisciplinary Reviews: Data Mining and Knowledge Discovery*, 3(1):12–27, 2013.

Rosé, C., Y. C. Wang, Y. Cui, J. Arguello, K. Stegmann, A. Weinberger, and F. Fischer. Analyzing collaborative learning processes automatically: Exploiting the advances of computational linguistics in computer-supported collaborative learning. *International Journal of Computer-Supported Collaborative Learning*, 3(3):237–271, 2008.

Ross, S. M., H. Kalman, J. E. Kemp, and G. R. Morrison. *Designing Effective Instruction*, 5th ed. John Wiley & Sons, Hoboken, NJ, 2007.

Schatz, S., D. T. Fautua, J. Stodd, and E. A. Reitz. The changing face of military learning. *Journal of Military Learning*, 1(April 2017):78–91, 2017.

Scheffel, M., H. Drachsler, S. Stoyanov, and M. Specht. Quality indicators for learning analytics. *Journal of Educational Technology & Society*, 17(4):124–140, 2014.

Schutt, R., and C. O'Neil. *Doing Data Science*. O'reilly Media, Sebastopol, CA, 2013.

Senge, P. M. *The Fifth Discipline: The Art and Practice of the Learning Organization*. A Currency book. Doubleday/Currency, 2006.

Shum, S. B. Learning analytics. Technical report, UNESCO Institute for Information Technologies in Education, Moscow, Russia, 2012.

Siemens, G. Learning analytics: Envisioning a research discipline and a domain of practice. *2nd International Conference on Learning Analytics & Knowledge*, 2012.

Siemens, G., and R. S. Baker. Learning analytics and educational data mining: Towards communication and collaboration. In *LAK'12 Proceedings of the 2nd International Conference on Learning Analytics and Knowledge*, pp. 252–254, 2012.

Siemens, G., D. Gasevic, C. Haythornthwaite, S. Dawson, S. B. Shum, and R. Ferguson. Open learning analytics: An integrated & modularized platform. *Knowledge Creation Diffusion Utilization*, pp. 1–20, 2011.

Smith, V. C., A. Lange, and D. R. Huston. Predictive modeling to forecast student outcomes and drive effective interventions in online community college courses. *Journal of Asynchronous Learning Network*, 16(3):51–61, 2012.

Sottilare, R. (Ed.). *Proceedings of the 5th Annual GIFT Users Symposium (GIFTSym5)*. U.S. Army Research Laboratory Human Research & Engineering Directorate, Orlando, FL, 2017.

Steinwart, I., and A. Christmann. *Support Vector Machines.* Information Science and Statistics. Springer, New York, 2008.

Tangient LLC. Wikispaces. https://www.wikispaces.com/. Accessed: September 13, 2017.

United States Army Program Executive Office for Enterprise Information Systems. Army Learning Management System. https://www.dls.army.mil/ALMS.html, 2017.

United States Army Training Support Center. The Army Distributed Learning Program (TADLP). https://www.atsc.army.mil/TADLP/.

Valsamidis, S., S. Kontogiannis, I. Kazanidis, T. Theodosiou, and A. Karakos. A clustering methodology of web log data for learning management systems. *Education Technology & Society*, 15(2):154–167, 2012.

Wang, A. Y., M. H. Newlin, and T. L. Tucker. A discourse analysis of online classroom chats: Predictors of cyber-student performance. *Teaching of Psychology*, 28(3):222–226, 2001.

Wasserman, S., and K. Faust. *Social Network Analysis.* Cambridge University Press, New York, 2009.

Wong, L., and S. J Gerras. *Lying To Ourselves: Dishonesty in the Army Profession*, volume February. U.S. Army War College Press, Carlisle Barracks, PA, 2015.

Chapter 6

Data Analytics for Electric Power and Energy Applications

Aaron St. Leger

Contents

Introduction

A reliable supply of electrical energy is critical for military (operational energy) and civilian applications. In civilian applications, health and welfare can be significantly compromised without a stable supply of energy. Currently, the availability of electrical energy to individuals and businesses is so reliable that for most applications, continuous availability of energy is assumed. A consequence of this is that in instances when this assumption fails, for example, the 2003 blackout of the northeastern United States and Canada (Office of Electricity Delivery & Energy Reliability, 2004), the effects can be severe. More than 60,000 MW of electric load serving approximately 55 million people was dropped during that blackout. While power was restored to most areas within 2 days, during this time the blackout shut down businesses, public transportation, and communication networks. Various analyses estimated the economic impact of this blackout to be between $4.4 and $10 billion dollars (Electricity Consumers Resource Council, 2004). These estimates included the effects of lost wages and business operations, spoiling of food, and emergency service costs. While this event is more of an anomaly than a common occurrence, it is a good example of the importance of energy reliability.

Data analytics and decision making in civilian electric power and energy applications are usually tied to economic or other derived metrics. Conversely, in military applications, the metrics are not as distinct or clear-cut, as costs and benefits associated with military operations are not as straightforward. While it is very clear that the ability to conduct effective military operations can be compromised without a stable energy supply, the structure and operation of military power systems is unlike most civilian applications and has unique motivations toward ensuring a stable supply of energy. Generally speaking, for short-term military applications similar to a temporary forward operating base (FOB), energy availability is the main mission support priority, and energy cost is a secondary concern. For longer-term applications, for example, a permanent military base connected to a commercial power grid, energy costs and long-term energy security become more of a factor to ensure operations in the event of power grid outages. The use of energy in military applications is referred to as Operational Energy (OE), which is specifically defined as "the energy required for training, moving and sustaining military forces and weapons platforms for military operations" (Office of the Assistant Secretary of Defense for Energy, Installations, and Environment). This is a broad definition that applies to many different systems including vehicles, ships, aircraft, and military bases (enduring and non-enduring). Within this framework, OE can be segmented into three distinctive categories of which the electric power systems, energy supply requirements, and data analytics have dissimilarities:

1. Platform power systems
2. Installation power systems (enduring bases)
3. Tactical power systems (non-enduring bases)

Platform power systems refer to a stable platform that includes an engineered power system to provide for the electric power and energy requirements. The only external requirement is a primary energy source (e.g., fossil fuels, wind, and solar radiation) that can be converted within the platform to electrical energy. Aircraft, vehicles, and ships are some examples of platform power systems. Installation power systems (IPS) refer to permanent military bases that are connected to the power grid and may have some generation on site. Last, tactical power systems (TPS) are those that supply electrical energy to non-enduring bases. One such example is a temporary FOB planned to operate for one year. Due to varying sizes, the non-enduring nature, and temporal and logistical considerations, a wide variety of power systems exist within this framework. This includes spot generation with diesel generators, direct power grid connection with backup generators, or a central power plant with an engineered electrical distribution system. Grid interconnection is typically favored, assuming there is a stable grid connection available. In lieu of this, depending on the size of the base and expected timeframe of base operation, spot generation or central power generation would be considered. Small, short-term bases will favor spot generation while larger, longer-term bases will favor centralized generation. However, predicting future requirements in FOBs is quite difficult as the military mission parameters will drive this.

Data analytics come into play on all categories of power systems, but in differing manners. For example, the cost of energy (economics) is an important issue for enduring bases. Additionally, given the long-term, stable nature of the bases, long-term data acquisition and analysis can drive decision making. Contrariwise, long-term data acquisition is not feasible nor relevant for short-term non-enduring bases. However, different data and analytics can help vastly improve the performance of the power systems. Last, platform power systems are typically engineered and built without much change until the next platform iteration. Initial design specifications and requirements will drive the power system design. Lacking clear pricing signals for many OE applications, energy policy targeted at military applications has been a driver of change. Furthermore, data analytics has been an integral component toward implementing operational energy policy.

Operational Energy Policy

Historically, the United States Military has treated energy as a commodity that would be readily available at all times, regardless of cost or manpower requirements. The primary focus was on supplying the energy requirements. Conservation, efficient use, and cost were secondary concerns. Recently, the supply of energy has become viewed as a security issue for military and civilian applications, and the Department of Defense (DoD) issued the first-ever Operational Energy Strategy in 2011 (Department of Defense, 2011). The overarching goal of this strategy was to guide the DoD to better utilize energy resources to support DoD missions, lower risk to Warfighters, and reduce energy expenditures. This policy not only provided specific guidance and goals for operational energy, but provided motivation and funding to help achieve these goals. Three distinctive high-level goals were identified in the DoD OE Strategy:

1. Reduce the demand for energy in military operations
2. Expand and secure the supply of energy to military operations
3. Build energy security into the future force

A clear strategic effort to reduce demand for energy in military operations was a key motivator toward improving energy efficiency. This was noted as the most immediate priority of the three strategic goals. Reducing energy demand yields significant cost savings, lowers fuel requirements (which improves energy security), and saves soldiers' lives. In tactical environments, fuel is the primary source of energy for vehicles and electricity generation. Costs associated with operating fuel convoys (labor, equipment, and force protection) can increase the cost of fuel significantly. At the time of this writing, the Defense Logistics Agency reports the commodity price of regular diesel fuel as $1.88 per gallon. However, the fully burdened cost of fuel (FBCF) has been shown to be as high as $600 per gallon in remote outposts (Dimotakis et al., 2006) and around $15 per gallon at more established locations (Eichenberg et al., 2016). More than 3000 Army personnel and civilian contractors were wounded or killed in action from attacks on fuel and water supply convoys from 2003 to 2007 in Iraq and Afghanistan (Army Environmental Policy Institute, 2009). From an installation power system perspective, reducing energy demand leads directly to cost savings. One of the initial subtasks in this effort was to incorporate and/or improve metering and OE consumption monitoring to help drive projects and benchmark results. The next component to the strategy was to expand and secure the energy supply.

Tactical power and platform power applications largely rely on a single source of energy supply: petroleum-based fuels. Diversifying the source of energy through alternative energy sources and alternative fuels was priority two in this strategic effort. The last goal of the effort is longer term. Initial energy demand reductions provided initial motivation and significant improvements to operations. However, it was anticipated that there would be a clear limit to the reductions obtainable. This strategic goal was designed to take a holistic view of military operations, to include energy supply logistics, acquisition of equipment, design and operation of vehicle platforms, and so on, and to ensure that energy supply to support these operations is considered up front, which historically has not been the case.

Significant progress has been made based on these goals, and an updated OE Strategy for 2016 has been published by the DoD (2015) based on progress and lessons learned from the first OE Strategy. One of the key requirements toward achieving these goals was the acquisition and analysis of appropriate data. Data obtained initially serves as a baseline measurement and provides inputs for analysis to develop and fund OE projects. Additionally, subsequent data collection and analysis allows for benchmarking results, evaluation of goals and progress, and provides input toward future projects.

Data Collection

Each of the three outlined OE categories have differences in their electrical energy requirements, production and delivery of energy, and timescale of operations. As a result, despite some commonalities, the data collection and data analytics applied to each category are unique.

Platform Power System Data Collection

Platform power systems are designed, built, and tested based on specifications and tests to validate performance prior to deployment. However, performance in these initial tests does not directly translate to real-world conditions due to a variety of factors, such as frequency of use, environmental conditions, and degradation of equipment. Because of the closed nature of these platforms, data collection needs to be customized and built within the initial design or retrofitted into existing platforms.

Installation Power System Data Collection

In the past, the information for installation power systems was primarily limited to information related to electrical energy billing from utilities and a small number of meters located within installations. Energy consumption over a billing period and peak electrical demand are commonly reported in electricity bills. Implementation of a robust metering system at all installations was accomplished in large part due to the 2011 DoD OE strategy. The DoD has been aggressively pursuing energy metering of individual buildings on installations.

Supported by policy, 100% of buildings above a threshold size have been metered for electricity, 78% for natural gas, 40% for water, and 100% for steam by the end of 2015. A goal for 100% of natural gas was set for 2016. This metering system includes meters installed on individual buildings and incorporates some of these meters into an automated metering system (AMS) to automate the reading and archiving of the data. The desired end state is to have all meters integrated into an AMS. The most advanced meters monitor numerous variables, perform some basic processing of data to obtain derived quantities, log data locally, and allow for remote communication to an AMS to view and archive data. A snapshot of an advanced electric power and energy meter monitoring a building is shown in Figure 6.1. These readings are taken at specified intervals (e.g., 5 minutes) and logged internally in the meter. The AMS communicates with the meter directly, requests readings at a user-specified interval, and archives the data within a database. Measurements include voltage (volts) and current (amperes) magnitudes for each of the three phases: current phase angle (degrees) and power factor of each phase, average demand (Watts) over the reporting interval, and a running total of real and reactive energy consumption (kilowatt-hours and kilovolt-ampere reactive

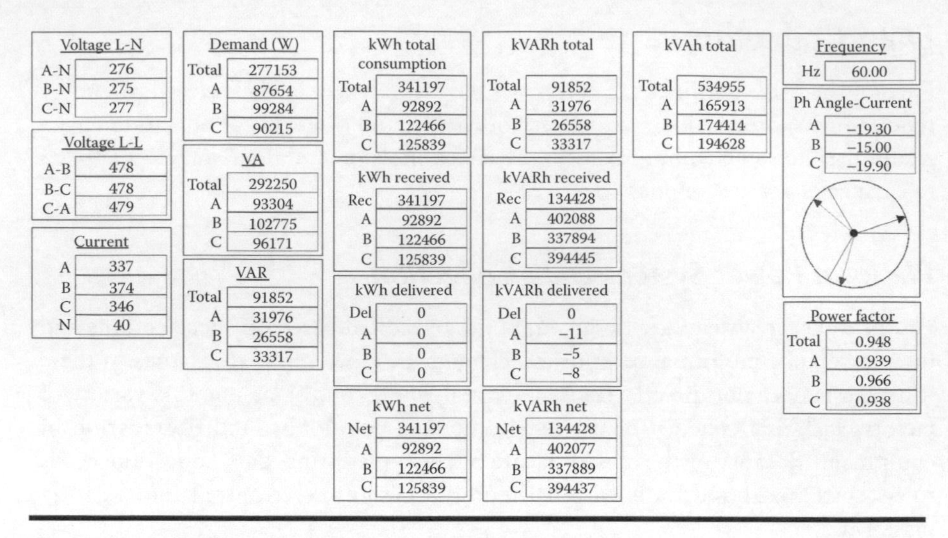

Figure 6.1 Reading from data-logging power meter.

hours, respectively). Similar meters can also be installed at the point of generation (e.g., solar photovoltaic system) within an installation to track energy generation. These meters are primarily designed to track energy consumption over time and sample data on the order of minutes. Phasor measurement units (PMUs) are a newer technology that allows for sampling voltage and current phasors up to 60 times per second in addition to other derived quantities. At the time of this writing, implementation of PMUs and the required communication infrastructure for transmitting data is cost-prohibitive at the installation level. However, there are efforts to develop low-cost PMUs suitable for application in distribution systems and large installations. This technology would enable more detailed monitoring and control of installation-level power systems.

Tactical Power System Data Collection

Tactical power systems can vary in size and complexity. Small and/or short-term outposts rely mostly on spot generation without much data collection in regard to electrical load. Fuel consumption is tracked to support fuel resupply. Larger, more permanent outposts will have prime power plants and electrical distribution systems installed and operated by trained personnel. Data collection and monitoring of this system is similar to that of installation power and has increased to support the OE strategy. However, data is not as widely available, and a standardized approach for data collection has not been deployed for TPS. Research and development on hybridized power systems, systems with two or more dissimilar energy sources, and tactical microgrids are providing more detailed data, but are not widely deployed at this time.

Operational Energy Applications

Platform Power Applications

Platform power systems can be improved through one of two options: improving operation and performance of existing platforms, or design platform power systems with a holistic approach toward energy. An example of the former is the Shipboard Energy Dashboard developed for the Navy. This system provides a real-time assessment of power and energy usage, reports this data via a graphical user interface onboard the ship, and archives data for further analysis. Operators can make decisions in real time to improve energy efficiency and operations, and archived data can be used for efficiency improvements and benchmarking. The concept of energy key performance parameters (KPP) has enabled an energy-focused approach to platform development (Bohwagner, 2013).

KPP is a mandatory analytical tool used to holistically view energy requirements for platforms in military applications. As discussed in (Bohwagner, 2013), a pure energy-efficiency approach is not sufficient, as an increase in energy consumption may coincide with an increase in capability that is worthwhile. The KPP defines the energy requirements of a platform by explicitly assessing the supply chain capacity and the energy demands based on expected operation, and, additionally, provides more standard metrics such as miles per gallon on a vehicle. The "cost" in energy versus the benefits gained in platform capability can be scrutinized and specified early in the development cycle using KPP. KPP is also a valuable tool for the planning and execution of energy supply once platforms are deployed.

Installation Power Applications

The primary objective of the 2011 DoD OE Strategy was to reduce the demand for energy. This was supported by the Energy Independence and Security Act (EISA) passed in 2007, which set a specific goal to reduce energy demand 30% by 2015. The 30% reduction was targeted against an aggregate energy baseline established in 2003. The progress of this effort, based on data obtained from the 2015 and 2016 DoD Annual Energy Management Reports (Department of Defense, 2016, 2017), is shown in Figure 6.2. Data is displayed individually for the Air Force, Army, and Department of the Navy (DoN). Total energy intensity reduction for the DoD is also shown. The metric used for measuring energy consumption and progress toward demand reduction is energy intensity (quantified in british thermal units [BTUs] per gross square foot of facility space). Scaling the energy consumption by facility space ensures that expansion or contraction of facility space will not yield an increase or decrease of energy consumption in the metric, which would lead to false signals in regard to energy reduction.

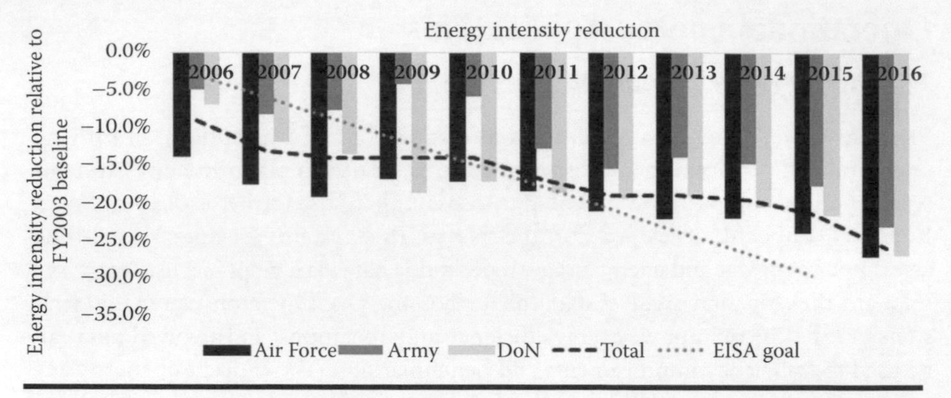

Figure 6.2 DoD energy intensity reduction.

The primary method of reducing the demand of installation energy was investment in efficiency and conservation projects on installations (Department of Defense, 2016). Examples of these projects include energy-efficient retrofits of lighting, heating, ventilation, and air-conditioning (HVAC) systems and building envelope improvements (insulation, windows). These efforts were funded through a DoD energy conservation investment program (ECIP) and performance-based contracts that showed financial gain through efficiency-improving projects. Proposed ECIP projects are evaluated and ranked by a savings to investment ratio. As a result, the program has favored smaller projects with rapid payback in regards to cost savings through energy reduction. While the 30% demand reduction by 2015 was not met, an overall reduction of 26.3% from the 2003 baseline has been achieved by 2016. Progress toward this initiative has slowed as most of the simple, cost-beneficial energy conservation efforts have already been implemented and resulted in the aggressive energy intensity reduction prior to 2012. Despite consistent ECIP funding, the rate of energy intensity reduction has slowed beyond 2012. Energy intensity reduction measures are becoming more expensive. Increasing ECIP funding, implementing more advanced meters, AMS, and energy monitoring systems (EMS), which enable meaningful analysis of metered data and more novel energy reduction projects, are paths forward for further gains and achieving the 30% reduction goal.

Advanced metering at the building level provides significantly more data at a higher fidelity as compared to the aggregate data from the 2003 energy baseline. For example, advanced electric meters monitor power and energy consumption at discrete time intervals, for example five minutes, which provides demand profiles on a per building basis. This data can then be processed, aggregated, and so on, via an EMS. The temporal aspect of the data, showing how power and energy demands vary over time, allows for the construction of demand profiles. An example demand profile is shown in Figure 6.3. Applications to reduce energy consumption and reduce energy costs are derived from this data.

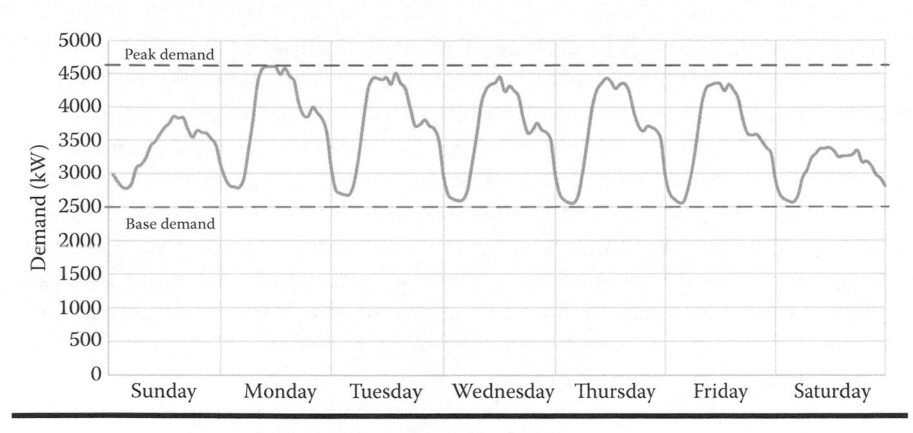

Figure 6.3 Example installation demand profile.

The same metric used in Figure 6.2, energy intensity, can be calculated for each building (per meter) as opposed to each installation. Additionally, with building demand profiles being directly measured, energy intensity over time can also be quantified and related to installation operations. This data permits detailed analysis of each and every structure on an installation and enables targeted efficiency improvements. For example, structures with higher energy intensities can be prioritized for efficiency improvements. The temporal aspect of the data allows for a more thorough analysis of how and when energy is being consumed and helps to identify inefficient use of structures. High energy intensity during non-work hours or in unoccupied buildings can be identified and corrected. This could occur if HVAC is running while a building is unoccupied. As can be seen in Figure 6.3, energy demand is higher during the week (Monday through Friday) and lower during the weekend. Furthermore, the demand profiles are similar for each weekday, which would be expected if operations and environmental conditions were similar for each weekday. Conversely, Sunday shows an appreciably higher peak demand, and energy consumption, as compared to Saturday. This could be flagged for further investigation to see where this energy consumption occurs and why it occurs during the weekend. There could be a good reason for this difference, or there could be an opportunity for reducing peak demand and energy consumption. These are examples of data-driven analysis that will allow further reduction in energy demand. Data derived from demand profiles also permit reduction in energy costs.

Most installations are billed for electrical energy in a similar fashion to commercial and industrial customers. Billing is comprised of three components: service charge, energy charge, and demand charge. The service charge is based on fixed utility operations costs dispersed across the customer base. The other charges are based on energy consumed over the billing period ($/kWh) and peak demand ($/kW). Peak demand is defined as the average demand over a time interval (usually 15 minutes) over the billing period. Relating to the demand curve, energy consumed is the area under the curve and peak demand is the peak value of demand, which is noted on Figure 6.3. Given this billing structure, reducing the peak demand of an installation can reduce the cost of energy for an installation, even if the same amount of energy is consumed. Demand response programs are targeted toward reducing this peak demand. This can be implemented manually by monitoring installation energy demand in real time via an EMS or automated via a controller. For example, air conditioning can be cycled off for a period of time to reduce peak demand. These demand profiles are also useful for studying the impacts of local electric power generation to include renewable sources of energy, evaluation and tracking of net-zero energy goals (Army Net Zero Energy Program), and design of installation-level microgrids to increase energy security. A more specific example would be the cost-benefit analysis of on-site diesel generation to reduce the peak energy demand seen in Figure 6.3.

The peak demand for the profile in Figure 6.3 is 4612 kW. A 650-kW diesel generator could be employed to conservatively reduce the peak demand to 4000 kW. The result of peak shaving is shown in Figure 6.4. The peak demand, from the utility perspective, would be 4000 kW. The local diesel generator would be used Monday through Friday during peak demand periods to service all demand beyond 4000 kW. Evaluating the direct cost benefit of this example relies on the cost differential between the generator cost and the utility cost to service this peak demand. The utility cost at this installation, which includes demand and energy costs, is on average $0.09/kWh. However, given the demand-based billing structure, the savings from peak-shaving is often much higher than the

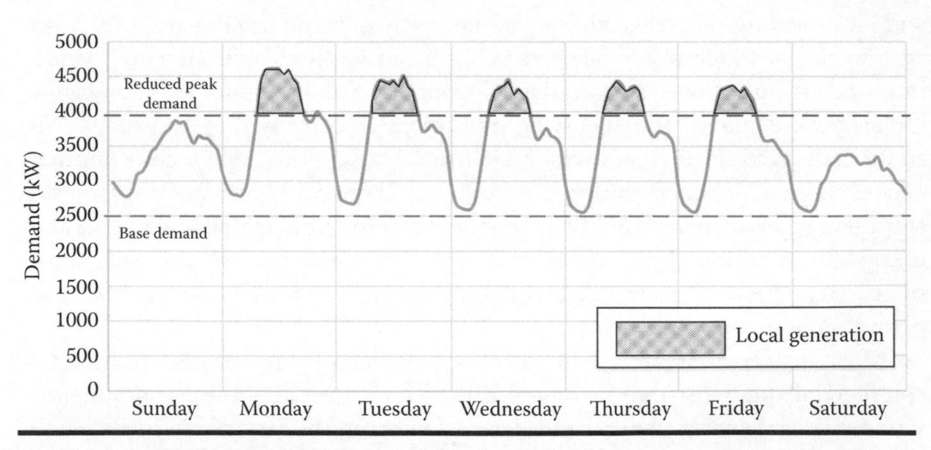

Figure 6.4 Peak-shaving example demand profile.

Table 6.1 Peak-Saving Cost Analysis

Energy		*Demand*		*Savings*	
Reduction (kWh)	63113	Peak Reduction (kW)	612	$/kWh	$0.24
Cost ($/kWh)	$0.06	Cost ($/kW)	$18.78	Total	$12,432.36
Savings	$946.70	Savings	$11,485.66		

average energy rate paid by a customer. Table 6.1 enumerates the savings for this example. Assuming a 4-week billing period with this demand profile, and a demand charge of $18.78 (average annual charge), reducing peak demand by 612 kW saves $11,485.66. The energy savings at $0.06/kWh (annual average rate) is $3,786.79. Overall, the savings per unit energy reduction from the utility is $0.24/kWh. This is approximately four times the average cost of energy purchased from the utility and provides a realistic opportunity to deploy local generation in a cost-beneficial manner. Further analysis on specific generator choice, operation and maintenance costs, peak demand/energy savings over an annual period, and so on, would need to be performed to more accurately quantify the cost-benefit of this project. Additionally, on-site generation contributes to net-zero energy goals and can provide a secondary purpose of a backup power source, increasing the energy security of the installation. While benefits of energy security and net-zero are difficult to quantify in terms of cost, they are important considerations for military applications.

Net-zero energy goals seek to generate as much energy within an installation as is consumed. From an electrical energy perspective, this is expected to be provided by a mix of renewable and alternative energy sources such as wind and solar photovoltaics as well as traditional generators driven by fossil fuels. In addition to local generation, installation-level microgrids are being investigated as a solution to energy security. Microgrids enable the installation power system to operate independently from the grid if required. By having the appropriate local generation and control capability, installations can operate critical buildings and services without electric grid connections for a prolonged period of time, ensuring the supply of energy and completion of mission and training requirements. Demand profiles allow for precise quantification of power and energy requirements for critical operations and will drive microgrid designs. Important design criteria for microgrids include serving peak demand, providing adequate energy over time, and maintaining base generation at all times. This problem is fairly straightforward if you install a fossil fuel–based generator that can serve the peak demand. However, from an energy security perspective, the supply of fuel becomes an issue.

Renewable sources of energy (e.g., solar and wind) are more desirable as they do not require a fuel source but are variable in their output based on wind and solar conditions. The optimal solution to this problem will be a mixture of fossil fuel–based and renewable generation coupled with demand-response controllers and energy storage. This will help meet electrical demand while minimizing fuel requirements. However, the technology in this field, as well as cost and capability, are rapidly changing, and projects must be analyzed on a case-by-case basis.

Tactical Power Applications

A study performed by MIT Lincoln Laboratory investigated performance of off-grid tactical power systems (TPS) (Van Broekhoven et al., 2014). This report noted that most diesel generators in tactical power applications were operating at only 10%–20% of rated load. This low load factor, which is defined as the ratio of average demand to the generator rating, increases maintenance requirements, significantly impacts generator efficiency, and increases fuel consumption per unit energy generated. Generally speaking, diesel generators should operate above 50% of their rated capacity to provide good efficiency and minimize maintenance requirements. One of the challenges associated with improving this performance is the lack of technical expertise in the field to set up, operate, and maintain TPS. Efforts to improve this are focusing on designing more efficient deployable power systems and providing more feedback to system operators in the field. A key component to designing more efficient deployable power systems is estimating the demand profiles.

A component-based approach to synthesizing FOB demand profiles, to provide data similar to that shown in Figure 6.3 for TPS, was introduced in Eichenberg et al. (2016). This fidelity of data is typically not recorded during system operation and can vary greatly based on location, FOB composition, mission requirements, and so on. Metering could be deployed to monitor an FOB and provide the data directly, but by the time the data is collected and analyzed, the FOB may not exist anymore or could have changed in size and/or operational scope, making this data and analysis much less useful. So, synthesizing the data a priori is a more viable and useful option. While predicting the temporal changes of FOB operations remains difficult, a priori analysis and design can yield good performance upon initial operation of the TPS. The approach for demand profile synthesis is to obtain a list of all equipment with electrical demand for a given FOB, environmental information for the location (temperature data), and operational information (number of soldiers, schedules, operational details). Demand profiles for each individual component are synthesized and aggregated to develop an estimated demand profile over the course of a year to capture seasonal and diurnal variations. Given a reasonable approximation of a demand profile, the generator loadout(s) and electrical distribution systems can be engineered a priori, provided to soldiers as a plug-and-play system that is already designed for the specific application, and enable increase in load factor and system efficiency. Additionally, the impacts of high-efficiency equipment and structures, and integration of renewable sources of energy, can be analyzed prior to deployment as shown in Eichenberg et al. (2016). Another application in TPS is integrating more advanced metering systems. While somewhat cost-prohibitive given the timescale of TPS operation, a basic low-cost EMS can provide meaningful real-time data and analysis to improve TPS operation and make educated decisions on TPS design changes as the size and operations of the FOB change over time.

A prototype EMS for a hybrid tactical power system (HTPS) containing a solar photovoltaic source and a diesel generator was developed at the United States Military Academy. A screenshot of the EMS dashboard is shown in Figure 6.5. This prototype provides real-time feedback on the operation of the HTPS and logs data for analysis. Electricity demand, photovoltaic energy production, and fuel consumption of the diesel generator are monitored in real time. Analysis of spot generator sizing is one example that this system provides. For this example, the generator load is monitored and averaged over discrete intervals of time (e.g., minutes) and logged. This data provides a detailed generator load profile as shown in Figure 6.6. This data is then processed to evaluate appropriate generator sizing in accordance with ISO 8528 standard (Reciprocating internal combustion engine driven alternating current generating sets, 2005). ISO 8528 classifies generators used in spot generation applications as prime power units, and the criterion for acceptable operating range is related to the generator power rating. Specifically, prime power generators should have a load factor of no more than 70% and can sustain an overload of 10% over nameplate rating limited to one hour over 12 operational hours not to exceed 25 hours per year of operation. Analysis of these metrics was performed on the generator load profile in Figure 6.6 for a 100-kW generator. Results are shown in Table 6.2.

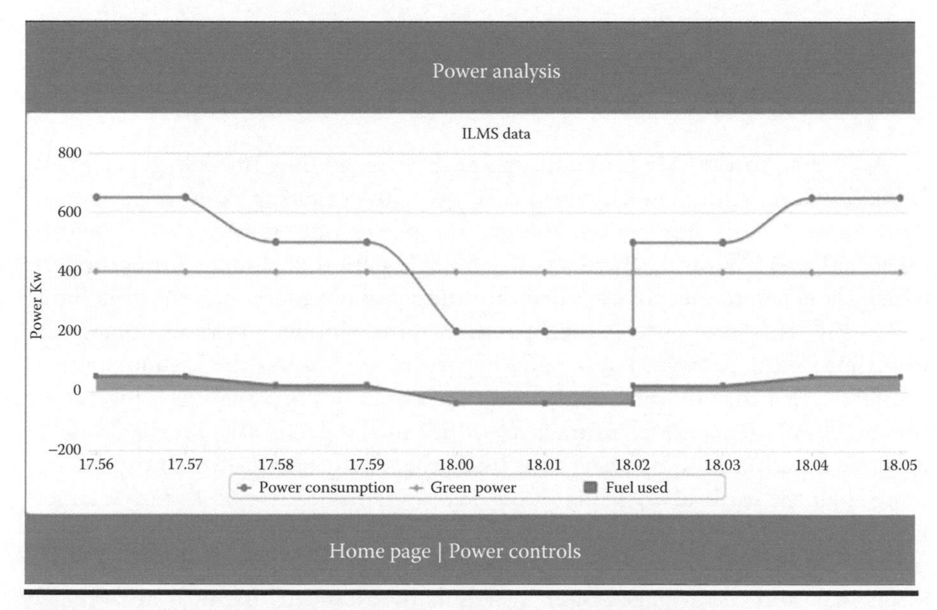

Figure 6.5 Prototype tactical power system energy dashboard.

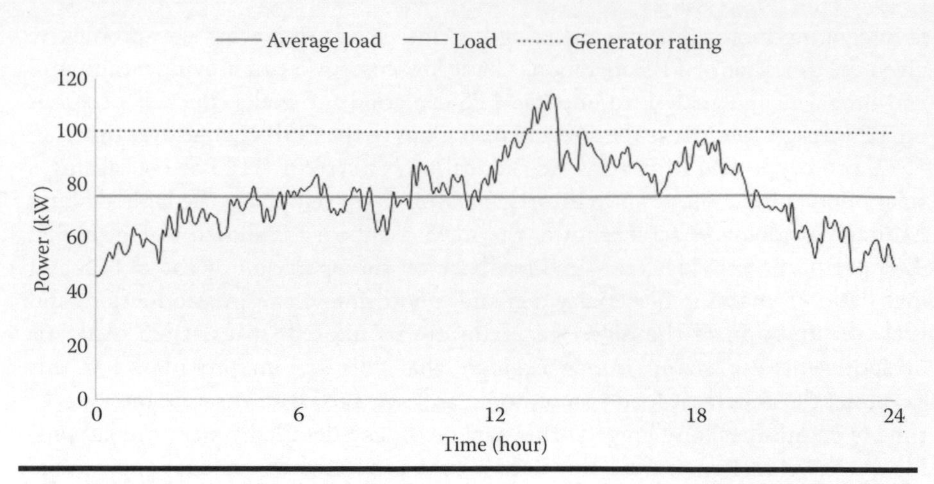

Figure 6.6 EMS generator monitoring example.

Table 6.2 EMS 100-kW Generator Monitoring Data Analysis

Metric	*Limit*	*Value*
Average power (kW)	70	76
Load factor (%)	70%	76%
Peak power (kW)	110	109
Total overload time (hours)	25	0.9
Continuous overload time (hours)	1	0.9

According to the ISO standard, the peak overload of a 100-kW generator is 110 kW and maximum average load is 70 kW (70% of rating). For this example, which spans a 24-hour period, the average power output was 76 kW, which exceeds the 70 kW limit. There was also a single period of time (~54 minutes) in which the generator output exceeded the rating, but not more than the peak limit of 110 kW. This is acceptable over this short-term window. However, total overload time would be tracked over a 365-day period to monitor the 25-hour annual overload time limit. In isolation and with extrapolation, this analysis is indicative that a 100-kW generator is insufficient for this load. Specifically, for the 24-hour example experiment shown here, it is likely that the generator would experience more than the limit of 25 hours of overload over the course of a year. A larger generator would be required. This type of monitoring system can notify an operator of the issue prior to generator damage. However, this is a short-term analysis of one day, and the average power is marginally exceeded. If this is an atypical peak load, for example an abnormally hot day with a high air-conditioning load, then this generator may fall within nominal operating constraints for most days and not exceed the annual overload time limit. Continuous analysis over time provides better insight into generator sizing and allows for upsizing and downsizing generation appropriately as mission energy requirements change. The system can also evaluate performance of the HTPS in terms of fuel savings related to the integration of a renewable source of energy. While still in the research and development stage, the goal is to have this type of system deployed with all TPS and allow operators to properly size generators and operate the system based on actual performance.

Conclusions

The advent of a unified DoD operational energy policy has driven improvements in operational energy over the past 5 years. The strategy focused on reducing energy demand, expanding and securing the supply of energy to military applications, and building energy security into future military applications. Obtaining and analyzing high-fidelity power and energy data has been imperative to the progress of these objectives. The data analytics and applications in electric power and energy summarized here have contributed to reducing the overall operational energy demand in 2014 for the United States Department of Defense by 30% from its peak in 2007 (Department of Defense, 2015). This 30% figure includes fuel reductions for vehicles. For electric power and energy applications, the initial efforts focused on establishing energy consumption baselines, data collection, and energy-efficiency programs. Large reductions in energy consumption via energy-efficiency programs were seen immediately but have been seeing diminishing returns. Moving forward, more detailed power and energy data that is now available is driving further improvements via microgrid design, establishing KPP for platform power systems, and integration of renewable energy sources. An updated 2016 DoD OE Strategy has recently been published to refine this strategy and continue these operational energy efforts.

References

Army Environmental Policy Institute, *Sustain the Mission Project: Casualty Factors for Fuel and Water Resupply Convoys Final Technical Report*, September 2009.

Army Net Zero Energy Program. Available: http://www.asaie.army.mil/Public/ES/netzero/.

A. Bohwagner, *An Overview of the Energy Key Performance Parameter* (*KPP*), December 9, 2013. Available: http://www.acq.osd.mil/eie/Downloads/OE/Energy%20KPP_12.

Department of Defense, *Annual Energy Management Report: Fiscal Year 2015*, Department of Defense, Washington, DC, June 2016.

Department of Defense, *Annual Energy Management Report: Fiscal Year 2016*, Department of Defense, Washington, DC, June 2017.

Department of Defense, *Department of Defense 2016 Operational Energy Strategy*, Department of Defense, Washington, DC, December 3, 2015.

Department of Defense, *Energy for the Warfighter: Operational Energy Strategy*, March 1, 2011.

P. Dimotakis, R. Grober, and N. Lewis, *Reducing DoD Fossil-Fuel Dependence*, The MITRE Corporation, McLean, VA, Technical Report Number JSR-06-135, September 2006.

N. Eichenberg, A. St. Leger, and J. Spruce, Impacts of thermally efficient structures and photovoltaic sources in military microgrids, *Proceedings of the 2016 Clemson University Power Systems Conference*, Clemson, SC, March 8–11, 2016.

Electricity Consumers Resource Council (ELCON), *The Economic Impacts of the August 2003 Blackout*, February 9, 2004. Available: http://www.elcon.org.

Office of the Assistant Secretary of Defense for Energy, Installations, and Environment, *Operational Energy*, 2017. Available: http://www.acq.osd.mil/eie/OE/OE_index.html.

Office of Electricity Delivery & Energy Reliability. U.S./Canada Power System Outage Task Force. *Final Report on the August 14, 2003 Blackout in the United States and Canada: Causes and Recommendations*, April 2004.

Reciprocating internal combustion engine driven alternating current generating sets: Part 1—Application, ratings and performance, ISO Standard 8528-1, 2005.

S.B. Van Broekhoven, E. Shields, S.V.T. Nguyen, E.R. Limpaecher, and C.M. Lamb, *Tactical Power Systems Study*, MIT Lincoln Laboratory, Lexington, MA, Technical Report 1181, May 2014.

Chapter 7

The Evolution of Environmental Data Analytics in Military Operations

Ralph O. Stoffler

Contents

Knowledge and consideration of environmental factors in battle often means the difference between a decisive victory and bitter defeat. Sun Tzu, a great Chinese general, in his fifth century BC *Art of War*, recognized this principle when he stated, "Know the Weather and Victory will be assured." These words ring true today in the United States Air Force. For example, the 557th Weather Wing logo (Figure 7.1) displays in Latin, "Choose the Weather for Battle." Throughout history, battlefield commanders have learned to respect and exploit the weather. In 1274, the Mongols attempted to invade Japan by sea, their vast fleet anchored off-shore as they prepared for the upcoming invasion. A powerful typhoon destroyed their overwhelming armed force, saving the Japanese from assured military defeat. The Mongols' lack of knowledge of this environmental factor was a key element in their defeat. As Napoleon marched across Europe, he achieved victory after victory until he reached Russia. Although he won key battles, the Russian counter-strategy deprived Napoleon's army of shelter and food, thus exposing his troops to the devastating effects of the Russian winter and forcing him to retreat. Russian knowledge of their own environment and their strategic utilization thereof was instrumental in Napoleon's defeat. In a more recent example, in 1944, Admiral William Halsey was involved in critical operations in the Pacific during World War II when he received word of an impending typhoon. At first, he decided to assume risk and not move his fleet out of harm's way. When he finally did direct his fleet to move, it was too late, and he lost significant elements of the United States Third Fleet to the typhoon.

Most military commanders recognize that environmental factors can jeopardize their operations. Predictive technologies are relatively recent applications that weren't available to the Mongols or Napoleon; but then, as today, commanders conducted a practice known as terrain walks or location surveys to assess the impacts of environmental conditions on their planned operations. In a terrain walk, high/low areas, soil conditions, rivers, pathways and other conditions are evaluated and considered. Seasonal patterns such as temperatures, flooding, or severity are considered also. Environmental conditions often repeat themselves in patterns, that is,

Figure 7.1 557th Weather Wing emblem.

the hurricane season, the sunspot cycle, the tornado season, and so on. These patterns or seasons are documented, and successful commanders incorporate them into their Risk Management Process.

The Mongols attempted a second invasion of Japan, and again their fleet was destroyed by a typhoon. They either failed to identify the typhoon season or opted to take a significant risk in planning their operation. In either case, their planned operation ended in disaster. Since long-range predictive technology is far from where we need it, we still prepare environmental assessments for all regions of the globe that military commanders can utilize in planning their operations. These assessments are used to choose the optimal environmental conditions for our force, if we can dictate the time an operation will be launched. If the time is not of our choosing, we attempt to find the silver lining in the environmental patterns by choosing weapons systems and means of employment that bring an advantage to our force.

Although military units have collected environmental information in order to characterize the environment, it wasn't until the twentieth century when we moved from the characterizing of the environment to the exploitation of environmental information. Traditional battles often took place in a small area over a short amount of time. As nations expanded their global interests, their militaries expanded to global operations. Technological innovations such as the radio provided the ability for global command and control. The battlefield grew in size, and the length of operations stretched across months versus days. Moreover, logistical resupply became even more critical in order to sustain operations. Two global wars helped accelerate this effort considerably, as nations wanted to minimize the impact of the environment on their forces and identify opportunities where the environmental factors could help defeat their enemies. Notable opportunistic examples are Germany's early efforts to understand the ionosphere to improve high-frequency communications with their U-Boat fleet, Eisenhower's weather decision for the D-Day invasion, and Germany's opting to launch the Ardennes Offensive in the foggy and low-ceiling winter of 1944–1945 to protect its tanks from Allied air power.

To get a better understanding of characterization and exploitation, let's walk through a common civilian example of characterizing and exploiting the environment. Formula 1 racing occurs in dry and wet track conditions. When rain occurs and the track becomes wet, drivers pull in for an unscheduled pit stop and change the tires. We characterized the weather as rain, the impact was a wet track, and the drivers mitigated the problem by changing tires. In exploitation, we take it to another level. Let's say we were planning a scheduled pit stop at 2 PM. If the rain starts slowly at 2:20 PM and the track is sufficiently wet by 2:35 PM to warrant wet tires, then, an unscheduled pit stop is needed at that time. If onboard fuel allows the 2 PM pit stop to be moved to 2:15 PM and we change the tires at that time, the driver will be slower for a few laps but will avoid the unscheduled pit stop, giving the team a significant advantage in time over the other teams. These extra seconds can mean the difference between victory and defeat. The same principles are applied in the military. If you are the aggressor, you plan your actions when the environment is in your favor. During a key Desert Storm operation, forecasted heavy cloud cover prevented opposition forces from visually spotting our approach aircraft; but, the forecast cloud breaks allowed our paratroopers to successfully land. Conversely, heavy sandstorms impacted our surveillance capabilities, which opposition forces exploited to move ground assets undetected.

In order to routinely achieve our goal of successfully characterizing and exploiting the environment, we have established a formal data analytics process that consists of five critical steps: collect, analyze, tailor, predict, and disseminate. These tasks are the core tasks of military environmental data analytics. This process has worked well through the years, but clearly technological improvements in computing power and communications have significantly improved our capabilities in each of these areas.

Collecting

Our collection process begins with gathering any environmental data that happens to be available via any means we can reasonably use. Dedicated communications circuits, high-frequency radio intercepts, the Defense Meteorological Satellites, and, of course, dedicated personnel that manually collect and document weather information are the primary sources. Generally, during the Cold War years, we believed little civilian environmental information would be available and potential adversaries would deny us access to their data. To circumvent this potential data loss, dedicated environmental satellites and terrestrial-based ground systems became the norm. Collected data was shared among military users and typically stored at the local level. This still occurs today. The end of the Cold War, the birth of Internet, and budget cuts have considerably changed the way we do business. To a great extent, the Cold War gave us a regional look and the conflict area was more predictable, allowing us to build a network of circuits and sensors and gain intelligence over potential conflict areas. We had significant familiarity of environmental data and its interaction with the terrain. Extensive rules of thumb and manual algorithms provided a definite edge over our adversaries. When the Fulda Gap and the Hof Corridor disappeared with the collapse of the Soviet Union, we became a truly global force, and information gathering became significantly more challenging. Technological innovations demanded a revision on how to process information. Dedicated communications circuits all but disappeared and were replaced by Internet-driven communications. High-frequency radio broadcasts became a thing of the past, and manual data sources were replaced by a sophisticated set of automated sensors. The end result: three times the amount of data. Where in the Cold War days you could often use one piece of paper to track a day's worth of data, now you can fill a sheet within minutes. We had to change the way we collected information.

Today, weather data is routinely collected at our primary military and central facilities. These facilities are linked in with military, civilian, and nontraditional sources of environmental information. They have the ability to provide this collected data to potential users across the globe for United States only and/or Coalition applications. These central repositories are the starting point for our data collection strategies. When a new mission is in the planning phase, we evaluate our existing data sources and determine if this is sufficient or if we need additional data. We also evaluate the reliability and assured access to data sources. Let's look at some examples.

A humanitarian relief mission where the military uses strategic airlift to a major international airport that has existing environmental sensors can generally be done within existing data sources. These types of missions are similar to their commercial counterparts already using the airports sensors. If the mission requires tactical airlift in the form of C-130s or helicopters to bare base (no existing sensors) sites, additional sensors will be required. Some military mission sets require unique sensors such as instantaneous versus average winds, for example. This requires unique mission sensors to measure that type of dataset. In some cases, existing civilian sensors are destroyed or can, as stated by treaty, only be used for non-combat missions. We would again either install our own sensors or leverage alternative data collections to obtain the data.

Coordination with the State Department, other nations, or commercial providers is used to secure access to existing datasets. Refresh rate of the data is also a key consideration. Many foreign sites update their data only once every 3 hours. A heavy airlift flow may require data updates every 20 minutes, driving the need for additional information. Accuracy and relevancy are also important aspects. Many observations across the globe are still manual and their recordings are often questionable or unrepresentative. Furthermore, some environmental datasets will report the worst condition and local conditions are often extremely variable. For example, in Iceland, one runway can be completely shut down by a blizzard while another one is clear for operations. Relevancy of the data is another key factor. Automatic sensors in many nations do not provide ceiling and visibility parameters, making the data less useful for a variety of military operations. Perhaps the most critical measure is access to the information. During combat operations, routinely available datasets are often disconnected for a variety of reasons. Amid the Desert Storm campaign, the primary dedicated environmental data feeds in Iraq were cut and existing sensors were destroyed. Data denial or falsification were issues experienced in Serbia and more recently in Syria. The Serbians deliberately sent false data hoping we would change our bombing strategy of Belgrade. Fortunately, verification with our satellite imagery prevented us from being duped by their efforts.

Once a complete picture of deployed data is available, the next step is to identify holes in the coverage. Critical needs are filled by deploying military equipment and personnel to close the gaps. When necessary, simulations are run to show ideal placement of collection points and what the potential improvement to our prediction capabilities would be if those additional sensors are used. We have introduced automated sensors linked with satellite communications to enhance our collections. This reduces our manpower footprint and moves more of our personnel out of harm's way. Smaller sensors allow deployment of equipment via smaller aircraft (i.e., helicopters) and increase data volume and refresh rates. Perhaps the most significant change, however, is where the data is collected. Today, central repositories are the location of choice. This enables faster processing, reduced equipment cost, and availability to all users simultaneously.

One of the most critical unplanned consequences of this technological revolution is cybersecurity. In the days of dedicated lines and manual data collections, our systems were secure. Today it is a far different picture. With a global network that routinely collects information from national and international sources, cyber vigilance requires the investment of significant resources. We continue to improve our capabilities through leveraging the latest remote maintenance and cybersecurity techniques. Shutting down compromised sensors is a top priority.

Analyzing

Satellites and ground-based sensors provide millions of environmental observations each day. If we add in nontraditional data sources (i.e., the Army's "Every Soldier Is a Sensor"), the volume of data runs into the petabytes each day. Prior to the use of automation and data processing, the majority of this analysis was done by hand, and to a great extent, we focused on a process called single station analysis. Essentially, relevant data for a small region was made available to locally based analysts. This data was plotted by hand on maps (Figure 7.2) and manually analyzed for key environmental features, such as weather fronts and areas of low ceilings and visibility. This process was cumbersome and often resulted in outdated information being presented to operators. Computers and software have improved this dramatically. In the manual days of our operation, new weather information generally became available once every hour with a maximum update every 20 minutes. The hand-analysis process allowed for updated weather analysis every 3 hours. Today, most environmental data is updated as changes take place and our systems can refresh every 2 minutes, making the latest data available to the operator and strategic planners as it occurs (Figure 7.3).

The most exciting area, however, is in articulating weather impacts to operations, platforms, and sensors. Once data is digitized, it can be presented in different ways and truly tailored to operational needs. Yesterday, an analyst could present a hand-drawn chart that depicted areas of low ceilings over an operational area. Today, that same analyst can present a route of flight showing ceilings that correspond to the unique aircraft and pilot limitations to accomplish the mission. We have identified over 100,000 environmental considerations categorized into 168 environmental parameters, all of which can be presented to the operator based on mission requirements.

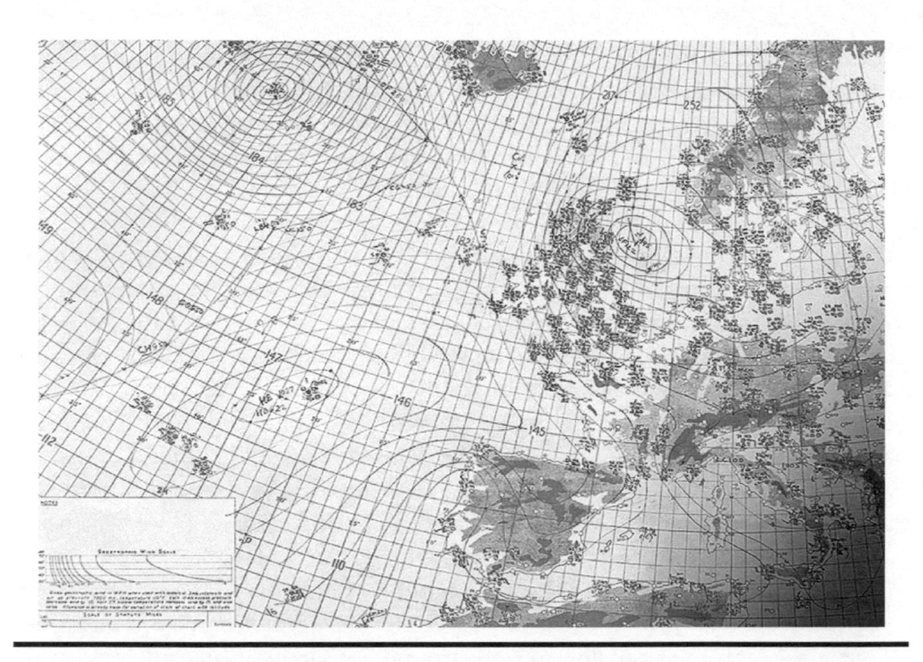

Figure 7.2 Hand-drawn analysis.

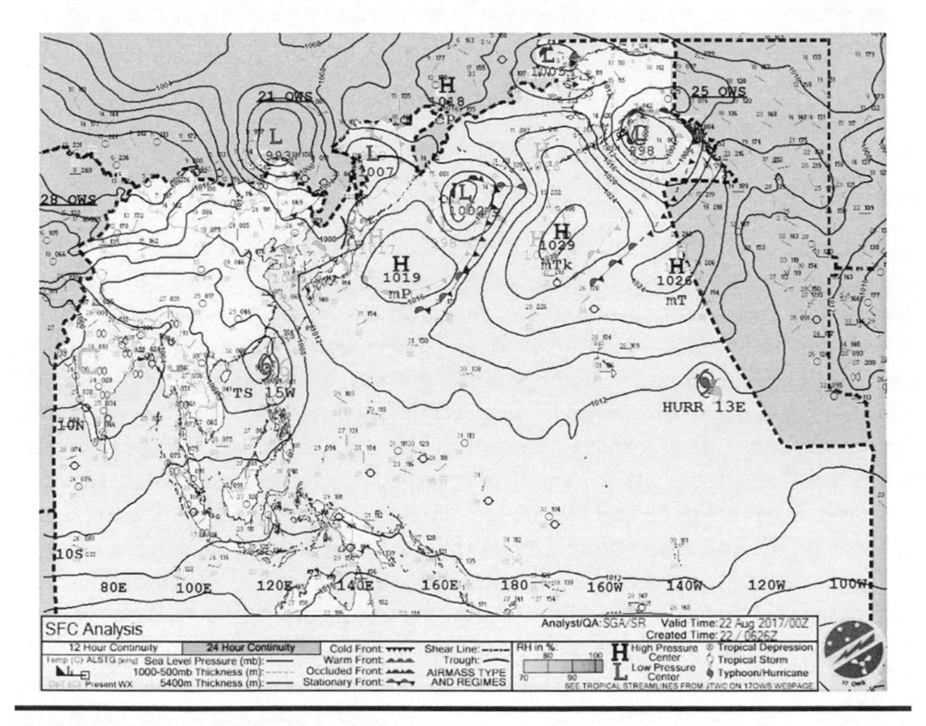

Figure 7.3 Automated analysis.

What are the drawbacks of this new wave of technology? As with most things, there are unintended consequences. Perhaps the most significant is that human analysts are losing their forecasting skills. The traditional process involved the human being, and in the process of hand plotting and analysis, the human analyst developed the ability to understand and predict the local environmental conditions. Today, we are served up automated information without understanding the underlying datasets, the algorithms that are being used, and the local terrain. The result is that human analysts increasingly bring less value to the process. We are going through a major effort to redirect human analysts' tasks to areas where they have the most value added. As an example, from one of our air bases, each morning the automated analysis shows clear skies and great flying conditions, yet the west end of the runway is covered in dense fog. Local review determined that land outside of the base was converted into a large rice field and when the fields flooded, the additional moisture contributed to the fog formation. To capture events that automation misses, we really need to understand the underlying algorithms and datasets so we can focus human analysts on where they add the most value to the process.

The second and more critical consideration is information overload. Operators are concerned about environmental information, but ultimately most want to be told what the problem is and how to work around it. Our automated analysis capabilities allow us to produce thousands of products a day, all of which we share with the operator. It's simply too much data, and we need to focus on what they need to know to get the mission accomplished. This brings us right to the next part of the process: tailoring.

Tailoring

The purpose of tailoring is to take our global dataset and our thousands of products and customize them to the mission at hand. Military leaders operate in a risk management framework and the environment is one of many factors they consider in the decision-making process. The operator's questions are always clear: Will the environment hamper my operation, and if so, can I mitigate it? In order to answer these questions, we need to fully understand the mission and then examine the dataset that is relevant to the mission profile. Key parameters of the mission profile often include the target objective, weapons platforms, weapons, movement and logistics issues, impacts on personnel, enemy impacts, route of flight, and many other details that may need to be considered. Much of this work may already have been done during the initial phase of the collecting process, but now as additional and final mission details emerge, we can understand which of the parameters we process will really apply to the mission at hand (Figure 7.4).

Multi-domain and overall large coalition operations pose particular challenges requiring key environmental impact decisions at the senior commander level,

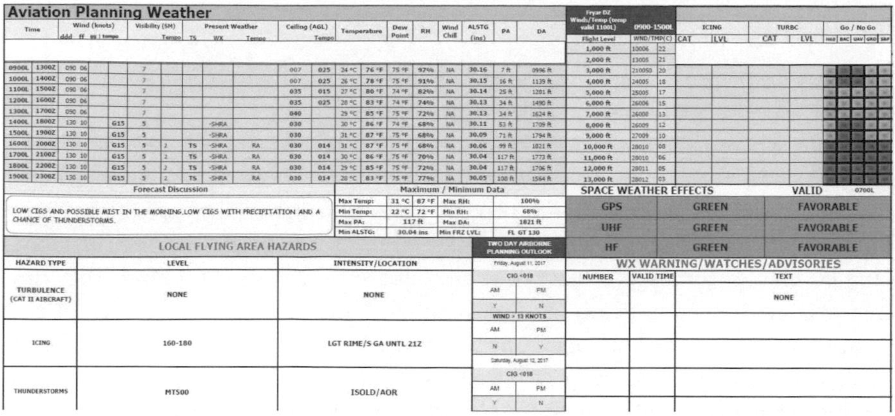

Aviation Planning Weather

Time		Wind (knots) ddd ff	gg	tempo	Visibility (SM)	Tempo	TS	WX	Tempo	Ceiling (AGL)	Tempo	Temperature		Dew Point	RH	Wind Chill	ALSTG (ins)	PA	DA
0900L	1300Z	090 06			7					007	025	24 °C	76 °F	75 °F	97%	NA	30.16	7 ft	0996 ft
1000L	1400Z	090 06			7					007	025	26 °C	78 °F	75 °F	91%	NA	30.15	16 ft	1139 ft
1100L	1500Z	090 06			7					035	015	27 °C	80 °F	74 °F	82%	NA	30.14	25 ft	1281 ft
1200L	1600Z	090 06			7					035	025	28 °C	83 °F	74 °F	74%	NA	30.13	34 ft	1490 ft
1300L	1700Z	090 06			7					040		29 °C	85 °F	75 °F	72%	NA	30.13	34 ft	1624 ft
1400L	1800Z	130 10		G15	5			-SHRA		030		30 °C	86 °F	74 °F	68%	NA	30.11	53 ft	1709 ft
1500L	1900Z	130 10		G15	5			-SHRA		030		31 °C	87 °F	75 °F	68%	NA	30.09	71 ft	1794 ft
1600L	2000Z	130 10		G15	5	2	TS	-SHRA	RA	030	014	31 °C	87 °F	75 °F	68%	NA	30.06	99 ft	1821 ft
1700L	2100Z	130 10		G15	5	2	TS	-SHRA	RA	030	014	30 °C	86 °F	75 °F	70%	NA	30.04	117 ft	1773 ft
1800L	2200Z	130 10		G15	5	2	TS	-SHRA	RA	030	014	29 °C	85 °F	75 °F	72%	NA	30.04	117 ft	1706 ft
1900L	2300Z	130 10		G15	5	2	TS	-SHRA	RA	030	014	28 °C	83 °F	75 °F	77%	NA	30.05	108 ft	1564 ft

Fryar DZ Winds/Temp (temp valid 1100L)	0900-1500L		ICING		TURBC		Go / No Go				
Flight Level	WND/TMP(C)		CAT	LVL	CAT	LVL	HGO	BAC	UAV	GRD	SBF
1,000 ft	10006	22									
2,000 ft	13005	21									
3,000 ft	210050	20									
4,000 ft	24005	18									
5,000 ft	25005	17									
6,000 ft	26006	15									
7,000 ft	26008	13									
8,000 ft	26009	12									
9,000 ft	27009	10									
10,000 ft	28010	08									
11,000 ft	28010	06									
12,000 ft	28011	05									
13,000 ft	28012	03									

Forecast Discussion

LOW CIGS AND POSSIBLE MIST IN THE MORNING.LOW CIGS WITH PRECIPITATION AND A CHANCE OF THUNDERSTORMS.

Maximum / Minimum Data

Max Temp:	31 °C	87 °F	Max RH:	100%
Min Temp:	22 °C	72 °F	Min RH:	68%
Max PA:	117 ft		Max DA:	1821 ft
Min ALSTG:	30.04 ins		Min FRZ LVL:	FL GT 130

SPACE WEATHER EFFECTS VALID 0700L

GPS	GREEN	FAVORABLE
UHF	GREEN	FAVORABLE
HF	GREEN	FAVORABLE

LOCAL FLYING AREA HAZARDS

HAZARD TYPE	LEVEL	INTENSITY/LOCATION
TURBULENCE (CAT II AIRCRAFT)	NONE	NONE
ICING	160-180	LGT RIME/S GA UNTL 21Z
THUNDERSTORMS	MT500	ISOLD/AOR

TWO DAY AIRBORNE PLANNING OUTLOOK

Friday, August 11, 2017	
CIG <018	
AM	PM
Y	N
WIND > 13 KNOTS	
AM	PM
N	Y
Saturday, August 12, 2017	
CIG <018	
AM	PM
Y	N

WX WARNING/WATCHES/ADVISORIES

NUMBER	VALID TIME	TEXT
		NONE

Figure 7.4 Mission planning product.

and the environmental analyst must understand these concerns. For example, a fighter pilot will be most concerned with impacts that directly hamper the mission from an aircraft flying in the kill box perspective, but a logistician will consider impacts to maintenance and resupply. The environmental assessment must cover all these areas to enable the senior commander to make an overall risk decision (Figure 7.5). Many of these issues were encountered during the Bosnian campaign. Freezing rain in Germany slowed ground resupply, heavy seas in the Adriatic Sea impacted sea lift, and dense fog in Bosnia hampered airlift operations. The heavy flooding of the Sava River slowed the ground force advance and severe weather over the Alps hampered air support. Fortunately, our ability to provide tailored environmental products aided senior commanders in their final risk decisions.

	Visibility	NGV Abs Humidity	Storm Surge	Ice/Snow Roads	Lightning	Turbulence	Winds	Heat Stress	Cold	Water Current
Bridges	1mi	G	G	R	G	G	30 Kt Gust	G	20F	G
Waterways	1mi	G	G	R	G	G	30 Kt Gust	G	20F	G
Railyards	1mi	G	G	R	G	G	30 Kt Gust	G	20F	G
Railroads	1mi	G	G	R	G	G	30 Kt Gust	G	20F	G
Airports	1mi	G	G	R	G	G	30 Kt Gust	G	20F	G
Ports	1mi	G	G	R	G	G	30 Kt Gust	G	20F	G
EM Ops Ctr	1mi	G	G	R	G	G	30 Kt Gust	G	20F	G
Personnel	1mi	G	G	R	G	G	30 Kt Gust	G	20F	G

Comments: Visibility will improve above 21miles by 1000L, winds gusts up to 30 knots.

Threshold Color Criteria Customized For Functional Operations Risks and Impacts

G	= No Weather or Water Impacts Expected
Y	= Marginal Conditions--Caution, Pay Attention for Updates in Case Conditions Change Worse Than Expected
R	= Operational Threshold Exceeded, Risk to Personnel Safety and Operations Effectiveness and/or Efficiency

Figure 7.5 Impact prediction product.

Predicting

Obviously, knowing the environmental impacts of yesterday and today are important, but what we really need to know is the future.

This is where technological advancements provide the greatest breakthrough. Environmental forecast models designed to cover the globe take a considerable amount of computing capability to run. 30 years ago, the military ran global predictions with a 60-km grid. To give you a better visualization, imagine the globe being covered by a mosquito net where the square in the net is 60 km by 60 km. Then overlay the net vertically 70 times and that image gives you an idea of the geographic scale in the horizontal and vertical levels (Figure 7.6). This is a lot of processing that needs to be done. Now add time to the equation. The overlays I just described give you a prediction for just 10 minutes, and the same process is redone for a 7-day forecast using 10-minute increments. That procedure was considered state-of-the-art in the days of the Cold War. Today, with more powerful computers, we have enhanced things considerably and the 60 km grid has improved to a 17 km grid with plans to reach 10 km grids soon. Since scalability in the battlespace is a critical warfighter need, there are efforts to process data down to the 1 km grid in key areas of interest. Vertical levels will also increase, allowing incorporation of space environmental data by using Whole Atmosphere Models that extend

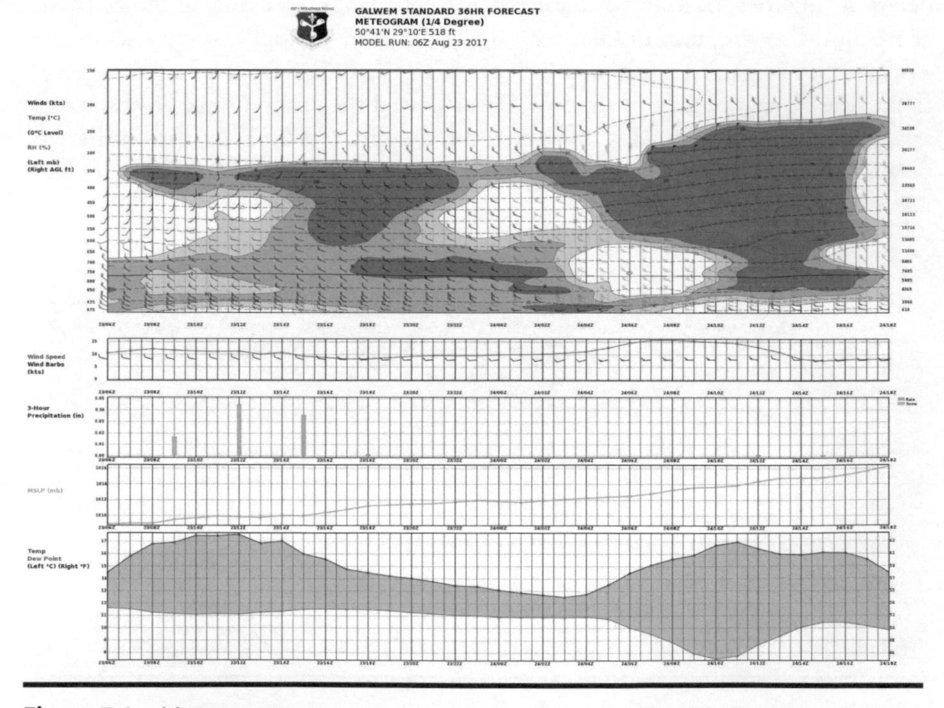

Figure 7.6 Meteogram.

potentially up to 600 km. Prediction length will go to 16 days, and a select number of parameters will be processed out to 45 days and longer. The environmental predictions are then transformed into tailored environmental impacts for specified operations.

Since our predictions aren't perfect yet, we have adopted the ensemble approach (advanced Monte Carlo simulations) to many parameters, and the concept is simple. When you solve a math problem, you expect the same answer every time you solve the same equation. If you have variables in your equation that can't be perfectly determined, you create uncertainty in your output. Ensembles compensate for this problem by recalculating the output multiple times with slight variations of the original variable input. If you rerun the output 40–100 times with different initial variables, you create an ensemble of possible solution sets. If all your runs provide close results, your probability of accuracy is high. Conversely, if the output is very diverse, then you may not have a handle on the situation and more reanalysis or data verification may be required. During the Bosnian campaign, our predictions and analysis pointed toward an intense low-pressure system. Actual conditions and other data did not support these conclusions. Detailed forensics of the data led us to remove an entire suite of sensors from the database as their barometers had fallen out of tolerance. Throughout the process, we apply horizontal consistency to ensure that products and data support logical, physics-based conclusions. For example, if products reflect good ground movement for tanks and yet precipitation is high, that would reflect poor horizontal consistency. With environmental parameters calculated over the time and distance required, we now run impact algorithms for missions, providing this output to planners for their decision-making process. Timing of the mission, approach path, and weapons platform selection are some of the areas that may be adjusted based on the environmental output. Likely probability of the event occurring is also a key consideration, especially in missions that are time dependent and can't be readily moved. During the Panama invasion, air refueling was critical to extend the range of our airlifters. Environmental prediction drove changes to air refueling locations to ensure our aircraft had sufficient fuel to reach the mission area and return. During a major REFORGER exercise in Europe, environmental prediction indicated heavy flooding and poor maneuverability, resulting in cancellation of the exercise. This action saved millions of dollars in possible maneuver equipment damage and lost training time. During the initial push into Bosnia, long-range environmental prediction clearly indicated that winter months would be the worst time to insert ground forces. Mission criticality and political pressure pushed ground forces to move in winter anyway. Deploying forces were equipped with engineer battalions to build mobile bridges to overcome the challenging floodwaters of the Sava River. In each case, the presentation of information revolved around the impact of environmental factors on the mission, enabling battlefield commanders to make the right risk decisions.

Disseminating

Information that doesn't arrive in time is wasted effort. Imagine hearing tornado sirens go off after your house has already been blown away. Effective and efficient communications are the lifeblood of the environmental data business. Most communications experts don't understand environmental data and are often surprised when our data hits their networks. For example, during a major Army Fifth Corps deployment, my supporting signal brigade assured me that our data would be no issue for the Corps' new tactical network. At precisely 0900, we turned on the data flow. By 0915, an excited signal officer was in our tent demanding we shut off the flow immediately, as the data had overwhelmed his system. In future exercises, the Corps would deploy two networks, one for the Corps and one for environmental data. The lesson learned is to understand your communications capabilities and only transmit what you need when you need it. Otherwise, you may jeopardize the entire network, and critical bits of information won't arrive in time. Most commanders utilize a variety of command and control systems that show information on tactical situations, and environmental information is incorporated into these displays. This allows a tactical analyst to develop different engagement strategies and quickly determine environmental impacts by overlaying the latest impact analysis mission profile. Although tactical communications are much improved, the more tactical you become, the smaller the communications lines are; hence, the need to send end user products versus raw unprocessed data forward. The bottom line is without good communications, you quickly revert to procedures from the Cold War era and rely on single station analysis to develop your environmental strategy.

The Military Toolbox

Successful implementation of the described processes requires well-trained personnel, high-quality sensors, high-speed communications, and state-of-the-art high-performance computers.

Personnel

We have a diverse mixture of military and civilian scientists and analysts to accomplish our environmental mission. Our analysts attend a unique DoD training school to learn the tools of the trade. These are non-degree-holding personnel that are trained over a 12-month period. These personnel are also trained in combat skills including parachute, scuba, and survival training. They have the ability to collect and provide predictive environmental data anywhere on the globe. Our scientists have university-level degrees, and we have internal capabilities to provide advanced academic training in tropospheric and space weather subjects. We employ meteorologists, hydrologists, oceanographers, space weather experts, and a variety of other specialties to get the job done. Our personnel are as good at analyzing environmental data as they are at firing their weapons or surviving in an austere environment.

High-Quality Sensors

Our sensors are modular and scalable. Some sensors are mobile (Figure 7.7); some are fixed. Some are on orbit or under the sea. In general, they measure atmospheric pressure, wind speed/direction, temperature, dew point, relative humidity, liquid precipitation, freezing precipitation, cloud height/coverage, visibility, present weather, runway visual range (RVR), lightning detection, weather radars, soil moisture, river flow, coronal mass ejections, sea surface winds, electron density, cloud visible and infrared, and geomagnetic fluctuations, as well as other parameters.

High-Speed Communications

We exchange data on unclassified and classified networks. Some dedicated lines are also in place. Satellites transfer data to and from austere locations anywhere on the globe.

High-Performance Computers

The DoD has a network of computers providing up to 26 petaflops of computer power. Only a small percentage of this is used for environmental processing, however.

Figure 7.7 Mobile weather unit in Bosnia.

Machine to Machine

The military conducts thousands of missions a day, all of which could be impacted by some environmental factor. Machine-to-machine transfers are key to support these operations. Our central processing node serves up processed gridded datasets to a number of command and control systems. These systems have built-in algorithms that run the datasets against the planned missions. Planners are given a quick read of how their missions would be impacted. Daily airlift operations are the most common area where this works very well. Planners enter a mission and the computer will quickly respond with a red, yellow, green coding. Red means high likelihood of "no-go." Green means go, and yellow generally means "too close to call." These missions are often reevaluated with human environmental analysts to more precisely make the go/no-go decision. Machine-to-machine transfers are the standard and will increasingly be part of our future.

Nontraditional Data

Generally, official environmental data is generated in proper formats for accredited and properly sited sensors. Automatic ingest of this data is often difficult and automatic authentication of the data is even more challenging. For special missions, human analysts review significant amounts of information to harvest additional knowledge. We are actively pursuing more automation in this area as more nontraditional weather data is being made available of areas of the globe where traditional data is sparse.

Artificial Intelligence and the Future

As an avid sci-fi reader, I remember James T. Kirk stating that Star Fleet would let it rain starting at 1420 and ending at 1450. I can assure you that we are a long way away from that, but the future will bring many changes. As we deploy self-driving cars, automated ships, and automated planes, environmental data needs to be part of the equation. In the future, these transportation devices will ingest gridded databases, exchange the environmental data they collect, and then adjust their route and speed to account for changes in the environment. The built-in artificial intelligence (AI) will improve our prediction capabilities by learning and adjusting. Just think: if all vehicles continuously updated the gridded database in the same area, the AI could adjust the predictive algorithms in that sector. With the addition of millions of sensors on thousands of vehicles, finer-scale algorithms and artificial intelligence will continue to improve our ability to operate in a changing environment.

As an experienced weather operator, I do want to address three major areas of concern for the future. First, as more organizations enter the automated transportation business, many are not considering the impact of the environment.

A small drone carrying military-related sensors could easily end up miles away if it's caught in a thunderstorm or any strong straight-line wind. Second, organizations that have equipped their vehicles with sensors to automate their travel often don't recognize that sensor performance is significantly impacted due to poor weather or solar activity. Thus, when you need your automation to perform at its best is when you might really be performing your worst. Finally, and perhaps most importantly, is real data versus generated data. As we operate in more austere places in the world, the need for real data and real sensors is vitally important. It is all too easy to simply generate data from the same predictive tools. Although scientists may see an impressive correlation between predictive analysis and generated data, from an operational perspective, we may be creating self-supporting data that is not representative or a realistic representation of existing conditions.

Environmental factors will continue to impact military operations, and the military will continue to lead the charge in applications and impacts of environmental information. As we continue to expand our space operations, leverage lasers weapons, implement hypersonic capabilities in the upper stratosphere, or implement self-driving trucks, we will develop new techniques and field new data collection sensors. These new sophisticated weapons systems mitigate or are unaffected by many traditional environmental issues, bringing new challenges, so we must continue to improve our understanding of the environment and how it impacts our systems and operations. We will continue to evolve our processes to achieve mission success, and at the same time, reduce risk and cost for global operations and improve overall understanding of the natural environment.

References

AFMAN 15-111, Surface Weather Observations
AFMAN 15-124, Meteorological Codes
AFMAN 15-129v1, Air and Space Weather Operations-Characterization
AFMAN 15-129v2, Air and Space Weather Operations-Exploitation
AFI 15-114, Weather Technical Readiness Evaluation
AFI 15-127, Weather Training
AFI 15-128, Air Force Weather Roles and Responsibilities
AFI 15-135v1, Special Operations Weather Training
AFI 15-135v2, Special Operations Weather Standardization and Evaluation
AFI 15-135v3, Special Operations Weather Team Operations
AFI 15-157, Weather Support and Services for the U.S. Army
AFI 15-182, Weather Functional Capability Management
AFPD 15-1, Weather Operations
AFVA15-136, Operational Weather Squadron Areas of Responsibility
AFH 11-203, Weather for Aircrews-Products and Services, April 11, 2017
Joint Pub 3-05 Doctrine for Joint Special Operations
Joint Pub 3-59 Meteorological and Oceanographic Operations

Chapter 8

Autoregressive Bayesian Networks for Information Validation and Amendment in Military Applications

Pablo Ibargüengoytia, Javier Herrera-Vega, Uriel A. García, L. Enrique Sucar, and Eduardo F. Morales

Contents

Introduction

Data integration and data cleansing are particularly relevant for military applications where trustable data can make a difference in life-threatening conditions. One of the aims in military data management is to have information superiority, which largely depends on the ability to have the right data at the right time. The right data not only means relevant information but also trustable information content that is used when the decision-making processes are made. There is a large number of military applications that depends on data obtained from different sources that need integration and cleansing, such as military surveillance and security domains.

An important and current application of data management is in the fight against terrorism. The key idea is to investigate and understand criminal behavior based on historical data extracted from pasts events. Large databases are being constructed in order to discover behavioral patterns that permit predicting future attacks.

In February 2017, the BBC published an article that relates the use of Facebook to detect terrorist activities using machine learning algorithms. When using Twitter, a database can be formed with different attributes like text content, user ID, other tagged individuals, timestamps, the language device used, and the user's location.

Caruso (2016) describes how Facebook and Twitter have special teams that detect terrorist activity on their social media and remove individuals or groups associated with terrorist activities. In 2016, Twitter suspended 125,000 accounts with links to ISIS.

Tuntun et al. (2017) utilize historical data of patterns followed by 150,000 terrorist attacks from 1970 to 2015. The authors comment that terrorists have learned that using social media risks detection and hence use alternatives. The analysis of databases includes 140,000 incidents considering approximately 75 features or attributes that characterize the criminal behavior.

This chapter proposes a novel and robust mechanism for information validation and amendment in databases where certainty in information is critical.

In data validation problems, missing data requires estimation and, if inaccurate, requires rectification. In both cases, anomaly detection and information reconstruction are necessary. Additionally, some contextual information may be needed, whether available from the same or a complementary data source. Given the critical importance that access to trustable data has to many industries, including the military, it is no surprise that data validation methods have been thoroughly researched across many of the more common issues: outliers (Abraham and Box, 1979; Balke, 1993; Hoo et al., 2002; Muirhead, 1986; Peng et al., 2012; Tsay, 1988; Walczak, 1995), sudden changes also referred to as innovation outliers (Abraham and Box, 1979; Balke, 1993; Marr and Hildreth, 1980; Muirhead, 1986; Tsay, 1988; Sato et al., 2006), rogue values (Herrera-Vega et al., 2018; Ibargüengoytia et al., 2006), and missing data (Dempster et al., 1977; Lamrini et al., 2011; Vagin and Fomina, 2011). Any of these deviations from regular data behavior may actually be due to exceptional circumstances and do not necessarily represent inaccurate information. In those cases, amendment is not necessary and the unusual data leads to actions such as raising an alert. But following the confirmation of inaccurate information, whatever the cause, the failing data has to be reconstructed, and for such purpose, it can be treated as missing.

Missing data refers to the problem where a gap in the information exists. Etiology is varied; it may be a missing sample, a necessity of out-of-boundaries inference, or the demand for a resampling, among others. In situations of missing information, interpolation/extrapolation techniques (Lancaster and Salkauskas, 1986) have dominated the scene. But interpolation is not the only option. When the available information comes from a single source with a certain temporal structure, then classical time-series modeling (Chatfield, 2004) such as Autoregressive Moving Average (ARMA) or Integrated Autoregressive Moving Average (ARIMA) has also been employed. Both approaches are appropriate for isolated data series and capitalize on within-variable information. When richer contextual information from a number of additional variables is available, a range of alternative techniques should be considered to exploit the complementary knowledge. These multivariate techniques may afford a reconstruction of the missing datum in terms of the nearest neighbor (Vagin and Fomina, 2011), self-organizing maps (Lamrini et al., 2011), or probabilistic graphic networks (Ibargüengoytia et al., 2013a), among others. These multivariate approaches have in common the exploitation of adjacent variables, often at the cost of ignoring any signal own information. In contrast with the rich literature available on different validation methods, the decision of when to choose one particular reconstruction strategy over another has been scarcely investigated. Little is known about when the dataset characteristics will favor the application of one technique over another. In such uncertain scenarios, a method that utilizes both sources of information, the signal-internal information and the related information present in the repository, may represent a compromise of the advantages of different approaches while alleviating the process of picking the best-suited data estimation approach.

Ideally, both the signal-internal information and the related information present in the repository should be taken into account for estimating the missing information, but this is an oversimplification. In every case, the weight given to the in-variable information, and the information from other variables should be reconsidered. In our research work preceding this chapter (Ibargüengoytia et al., 2013b), we showed how the performance of different data estimation approaches vary as the scenario of available information exhibits different properties, and more precisely, how this is dictated by the variable autoregressive order and its dependency on the additional known variables. In that previous work, a concomitant contribution was the proposal of a new model, the autoregressive Bayesian network (AR-BN), that balanced its output, aiming to perform robustly across a wide range of scenarios.*

Methods

In this section, a description of some available methods to complete databases is included. Later, a more complete description of the proposal of this chapter, namely the autoregressive Bayesian network, is included. A mathematical formulation of the problem of incomplete data can be found in Dempster et al. (1977).

Interpolation

Interpolation is a large family of models in which the value of a variable at some sampling location is estimated from neighbor observations. In general, the semantics of the sampling location is irrelevant for the model itself other than setting, which are the neighbor samples and how far they may be from the questioned sampled location. When the sampling location is within other observed locations, then these models are referred to as interpolation, and when beyond, then they are referred to as extrapolation. Traditionally, interpolation has been an easier guess than extrapolation.

Linear Interpolation

Perhaps the simplest interpolation approach is linear interpolation, by which, assuming that the function is locally linear, that is, the approximation using only the Jacobian system is considered sufficient, the value of the variable at the new location is given by the line crossing its two nearest-known neighbor observations. Let s_t† be the targeted new sampling location and discretely let s_{t-1} and s_{t+1} be the immediate previous and next neighbor locations on which observations X_{t-1} and X_{t+1} have already been made. The estimated value X_t is given by Equation 8.1:

$$X_t = X_{t-1} + (s_t - s_{t-1})\frac{X_{t+1} - X_{t-1}}{s_{t+1} - s_{t-1}} \tag{8.1}$$

* This chapter is an extension of the work presented by this group at the Eighth International Conference on Systems (ICONS) 2013 (Ibargüengoytia et al., 2013b), with emphasis on the description of the AR-BN model.

† Without any temporal semantics associated.

Spline Interpolation

Spline interpolation is a more sophisticated approach in which a differentiable curve is built in a piecewise manner between an arbitrary number of neighbor observations supporting the curve definition. The curve is expected to be differentiable at all supporting points and thus derivatives at those points also ought to be known. In the simplest case, when only two supporting points s_{t-1} and s_{t+1} and associated observations X_{t-1} and X_{t+1} are taken (the first derivatives at those points X'_{t-1} and X'_{t+1} can be estimated from using further subsequent neighbors), a third-order polynomial can be written as:

$$X_t = (1-h)X_{t-1} + hX_{t+1} + h(1-h)(a(1-h)+bh) \tag{8.2}$$

where:

$$h = \frac{s_t - s_{t-1}}{s_{t+1} - s_{t-1}}$$
$$a = X'_{t-1}(s_{t+1} - s_{t-1}) - (X_{t+1} - X_{t-1})$$
$$b = -X'_{t+1}(s_{t+1} - s_{t-1}) + (X_{t+1} - X_{t-1})$$

As the earlier system is underdetermined, second derivatives are commonly required to match the sampling locations to complete the system. Other polynomials can be constructed (de Boor, 1978), but they are beyond the scope of this chapter.

Autoregressive Models

Time-series analysis has traditionally focused on estimating future values of a variable that has (or is assumed to have) a certain dynamic, for example, general trend, seasonalities, and so on. Autoregressive models are perhaps the simplest of time-series models in which the next observation is derived from a linear combination of preceding observations. The general autoregressive model of order n—denoted AR(n)—is defined in Equation 8.3:

$$X_t = c + \sum_{i=1}^{n} \alpha_i X_{t-i} + \epsilon_t \tag{8.3}$$

where $\alpha_{i|i=1\ldots n}$ are the model parameters, c is a constant, and ϵ_t is noise. Note the assumed temporal semantics in contrast to interpolation, but beware that from an abstract point of view, it remains a discrete relation of ordering among the sampling locations. Hence, in an offline repository, "future" data may also be available and can be easily incorporated as in Equation 8.4:

$$X_t = c + \sum_{i=1}^{n} \alpha_i X_{t-i} + \sum_{j=1}^{m} \beta_j X_{t+j} + \epsilon_t \tag{8.4}$$

where $\alpha_{i|i=1\ldots n}$ and $\beta_{j|j=1\ldots m}$ are the AR(n, m) model parameters.

Probabilistic Modeling

Probabilistic models exploit the laws of probability to estimate the most likely outcome of some event, that is, location, defined over the probability space (Sucar, 2015). Like interpolation and autoregressive models, probabilistic models are also a large family that traditionally have been well-suited under uncertainty. Among them, a Bayesian network is a directed acyclic graph (DAG) representing the joint probability distribution of all variables in a domain (Pearl, 1988). Bayesian networks use the Bayes theorem that relates hypotheses and evidence and makes relations among variables graphically explicit, as can be seen in Figure 8.1. The graphical companion is not superfluous. The topology of the network conveys direct information about the dependency between the variables. The structure of the graph represents which variables are conditionally independent given another variable. The variable at the end of an arc end (variable E) is probabilistically dependent on the variable at the origin of the arc (variable H).

Thus, obtaining values of the evidence E, Bayesian networks calculate the probability of hypothesis H given the evidence. This corresponds to Bayes theorem where the computation of $P(H \mid E)$ is calculated using $P(E \mid H)$ and $P(H)$ as per Equation 8.5.

$$P(H \mid E) = \frac{P(E \mid H)P(H)}{P(E)} \tag{8.5}$$

The knowledge in a process using Bayesian networks can be represented with two elements: (1) the structure of the network, and (2) the parameters $P(E \mid H)$, that is, the conditional probability tables, and $P(H)$. These parameters are learned from observations when available. In the application of completing databases, the parameter $P(\text{missingvalues} \mid \text{relatedvalues})$ can be calculated using $P(\text{relatedvalues} \mid \text{missingvalues})$ if knowledge is available in these databases, that is, with complete historical dataset of the process.

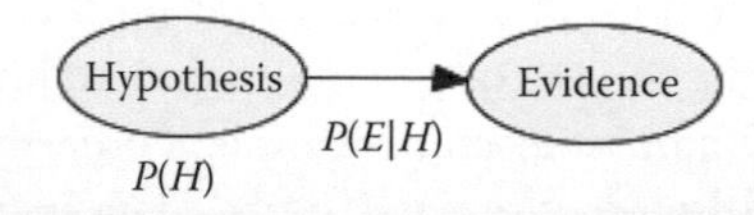

Figure 8.1 Elemental Bayesian network: Structure and parameters.

Knowledge about a system represented as a Bayesian network can be used to reason about the consequences of specific input data by what is called probabilistic reasoning. This consists of assigning a value to the input variables and propagating their effect through the network to update the probability of the hypothesis variables. The updating of the certainty measures is consistent with probability theory based on the application of Bayesian calculus and the dependencies represented in the network. Several algorithms have been proposed for this probability propagation (Pearl, 1988).

Bayesian networks can use historical data to acquire knowledge but may additionally assimilate domain experts' input. One of the advantages of using Bayesian networks is the three forms to acquire the required knowledge. First, the participation of human experts in the domain is quite effective; they can explain the dependencies and independencies between the variables and also may calculate the conditional probabilities. Second, there is a great variety of automatic-learning algorithms that utilize historical data to provide the structure and the conditional probabilities corresponding to the process where data was obtained. A combination of the previous two is the third approach, that is, using an automatic-learning algorithm that allows for the participation of human experts in the definition of the structure.

Dynamic Bayesian Networks

Plain Bayesian networks (BNs) consider only static situations of a domain. Time is not considered, and the calculation made on the hypothesis nodes consider only current values of the evidence nodes. The databases relevant in data validation in this chapter are usually time series where values are obtained in discrete intervals of time. Dynamic Bayesian networks (DBNs) are an attempt to add the temporal dimension into the BN model (Dean and Kanazawa, 1989; Mihajlovic and Petkovic, 2001). Often a DBN incorporates two models: an initial net B_0, learned using information at time 0, and the transition net $B_{\rightarrow}$, learned with the rest of the data as illustrated in Figure 8.2. Together, B_0 and $B_{\rightarrow}$ constitute the DBN (Koller and Friedman, 2009). An important assumption is made for DBNs: The process is assumed to be Markovian, that is, the future is conditionally independent of the past given the present. This assumption allows the DBN to use only the previous time-stage information in order to obtain the next stage.

A DBN can be unfolded over as many stages as necessary, and the horizontal structure can change from stage to stage. The resulting network is highly expressive but often unnecessarily complicated. Alternatives have been proposed to reduce this complexity like the Temporal Node Bayesian Networks (Herrera-Vega et al., 2018). In datasets arising from physical processes, statistical dependencies among variables can be expected to be stable across time. That is, if two variables, X and Y, are statistically dependent at time t_i, they will likely also be statistically dependent at time t_{i+j} for any arbitrary samples i and j, and similar reasoning can be made for independencies. This implies that the process is time-invariant, which can be

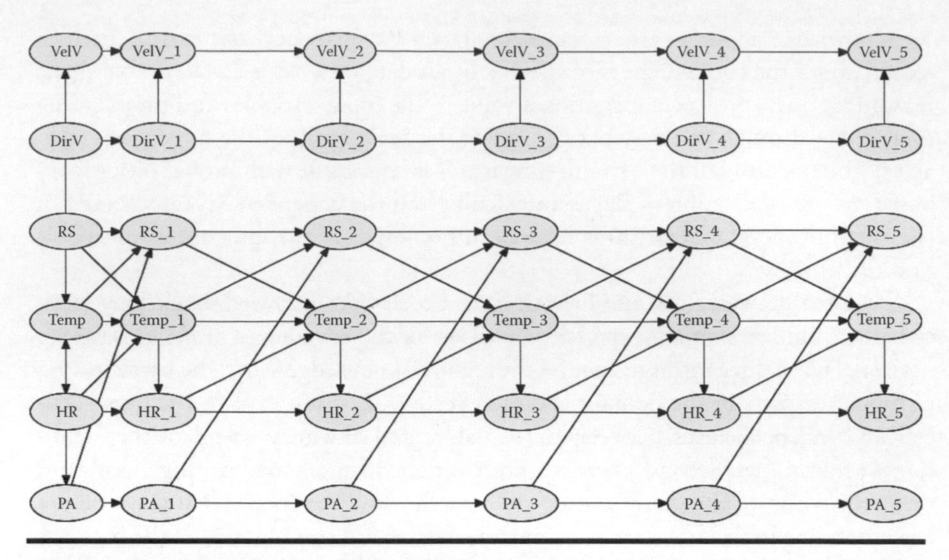

Figure 8.2 Dynamic Bayesian network.

exploited to simplify the model representation, as only the initial and transition networks are required.

Autoregressive Bayesian Networks (AR-BNs)

Autoregressive Bayesian networks are a simplified variant of DBNs. They incorporate the temporal dimension by observing time-shifted versions of the variables, whether past or future. Conceptually, they can be regarded as bringing an autoregressive model AR(n, m) to the BN domain.

Suppose a dataset with some dynamics of interest. Figure 8.3 illustrates the proposed probabilistic model. Variable X represents the variable to be estimated,

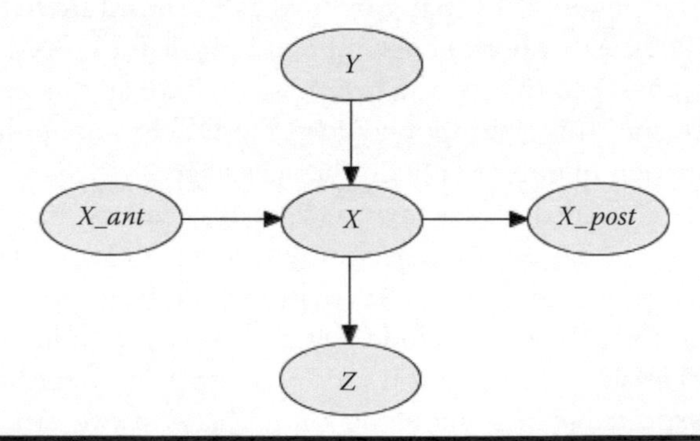

Figure 8.3 Dynamic probabilistic model proposed for data estimation. The structure can be enriched with other time-shifted versions of *X*, *Y*, and *Z* as appropriate.

variables Y and Z represent pieces of Bayesian network corresponding to all the related variables to X. *X_post* represents the value of variable X at the time $t + k$, and *X_ant* represents the value of variable X at the time $t - k$, although for simplicity in this chapter we will take $k = 1$.

This proposed model represents a dynamic model that provides accurate information for estimating the variable in two senses: first, by using related information identified by automatic-learning algorithms or experts in the domain, or both; and second, by using information of the previous and incoming values. This information includes the change rate of the variable according to the history of the signal.

In this approach, the horizontal (inter-stage) topology of the network is fixed. The persistency arcs among a variable and its shifted versions are enforced, whereas those between different variables at different stages are forbidden.

Estimating Missing Data from Incomplete Databases Using AR-BNs

The proposed procedure for estimating missing data from incomplete databases is in Algorithm 8.1. The first four steps build the model, and the last three propagate knowledge to estimate data holes.

Suppose a time series of three variables. Following Algorithm 8.1, a structure of the static version is obtained in step 3, as shown in Figure 8.4. In step 4, the network is extended and completes an AR-BN as shown in Figure 8.5.

Algorithm 8.1 Estimation of Missing Data

1: Obtain a complete dataset that includes information from the widest operational conditions of the target process.
2: Clean the outliers and discretize the dataset.
3: Utilize a learning algorithm that produces the static Bayesian network relating all the variables in the process. During the learning process, a complete train dataset with data from all variables is needed, as indicated in step 1.
4: Modify the static model to include previous and posterior values of every variable.
5: For all registers in an incomplete database, if one value is missing, instantiate the rest of the nodes in the model.
6: Propagate to obtain a posterior probability distribution of the missing value given the available evidence.
7: Return the estimated value with the value of the highest probability interval, or calculate the expected value of the probability distribution.

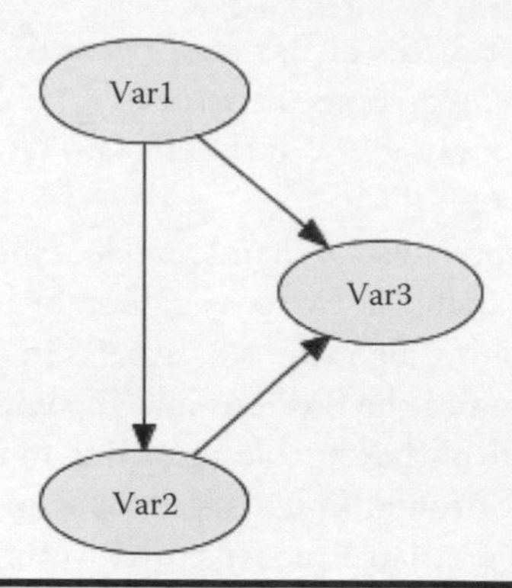

Figure 8.4 Static Bayesian network version of dataset 2.

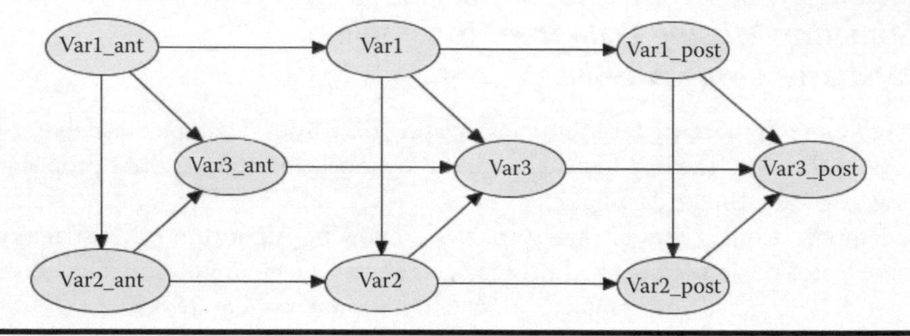

Figure 8.5 Autoregressive Bayesian network proposed for data estimation for dataset 2.

Notice that the creation of the structure, as in Figure 8.5, requires the calculation of parameters. These parameters are calculated, as mentioned in step 1 of Algorithm 8.1, using a complete dataset of the operational conditions of the target process.

Error Metrics for the Estimation of Missing Values

In order to establish the accuracy of the estimation of the missing values, the following error metrics were computed (Osman et al., 2001). Let E_i be the relative deviation of an estimated value x_i^{est} from an experimental value, x_i^{obs}:

$$E_i = \left[\frac{x_i^{obs} - x_i^{est}}{x_i^{obs}} \right] \times 100 i = 1, 2, \ldots, n \tag{8.6}$$

with n being the number of missing data.

- *Root Mean Square Error:*

$$E_{rms} = \left[\frac{1}{n} \sum_{i=1}^{n} E_i^2 \right]^{1/2} \tag{8.7}$$

- *Average Percent Relative Error:*

$$E_r = \frac{1}{n} \sum_{i=1}^{n} E_i \tag{8.8}$$

- *Average Absolute Percent Relative Error:*

$$E_a = \frac{1}{n} \sum_{i=1}^{n} | E_i | \tag{8.9}$$

- *Minimum and Maximum Absolute Percent Relative Error:*

$$E_{min} = min_{i=1}^{n} | E_i | \tag{8.10}$$

$$E_{max} = max_{i=1}^{n} | E_i | \tag{8.11}$$

These metrics will be used in the experiments conducted and discussed in the section "Experiments and Results."

Data Characterization

By analyzing the data, we can gain insight into the difficulty of validating and amending the datasets, as well as which method to complete the missing data could be more appropriate. To extract descriptive parameters that help us to know more about the behavior of the datasets, we have characterized them according to the methods described as follows:

- *Principal component analysis (PCA):* This technique, developed by Hotelling (1933), analyzes a dataset composed by intercorrelated variables and extracts the relevant information in the data. Then, this is represented as a set of orthogonal variables called principal components that correspond with the maximum variance. Data dimensionality is reduced by removing the components with less variance. The resulting variables represent the intrinsic dimensionality of the original dataset.
- *Intrinsic dimensionality:* The algorithm of Fukunaga and Olsen (1971) aims to look at local characteristics of the data distribution, establishing small subregions around each variable and, by the Karhunen-Love expansions for these subregions, determine the intrinsic dimensionality of the data.

- *Pearson correlation:* This is a well-known method to measure the linear dependence between two variables. Values +1 and −1 represent a linear dependence, and 0 indicates a nonlinear relation. The Pearson correlation coefficient is defined as follows:

$$r = \frac{\sum_{i=1}^{n}(x_i - \bar{x})(y_i - \bar{y})}{\sqrt{\sum_{i=1}^{n}(x_i - \bar{x})^2}\sqrt{\sum_{i=1}^{n}(y_i - \bar{y})^2}} \tag{8.12}$$

- *Akaike information criterion (AIC):* The AIC is a model selection method (Akaike, 1969) defined as:

$$AIC = 2 * N\log L + 2 * m \tag{8.13}$$

where m is the number of estimated parameters, and $N\log L$ is the log-likelihood. This method selects a model that minimizes the distance between the model and the truth. In autoregressive models, with Akaike's method, we select the order for which Equation 8.13 attains its minimum as a function of m (Shibata, 1976).

- *Kwiatkowski–Phillips–Schmidt–Shin (KPSS):* This method tests the null hypothesis that a time-series is trend-stationary against the alternative hypothesis that it is a nonstationary process (Kwiatkowski et al., 1992). Briefly, the KPSS breaks the time-series in three parts to construct a model (Equation 8.14) consisting of: a deterministic trend (βt), a random walk (r_t), and a stationary error (ε_t):

$$x_t = r_t + \beta t + \varepsilon \tag{8.14}$$

A least-squares regression is performed to fit the original data and the model. Finally, the data is considered stationary if the term (r_t) is constant.

Experiments and Results

This section describes the set of experiments conducted for the comparison of performance between different methods on different datasets.

Characterization of the Datasets for the Experiments

Simulations were carried out to reconstruct missing data from 2 different industrial datasets of different natures (variables have been enumerated for confidentiality).

The first dataset comprises 10 variables. It corresponds to a manufacturing process. The second dataset comprises 3 variables. It corresponds to an energy domain. Intrinsic dimensionality of the datasets as found by Principal Component Analysis (PCA) is 7 and 1 respectively (99% of variance included). For the dataset 2, the scale of one of the variables is 5 orders of magnitude larger than the remaining 2 variables. Hence, the global intrinsic dimensionality is perceived to be 1 by PCA, but local dimensionality of the dataset remains 3, which can be determined by Fukunaga and Olsen's algorithm (Chatfield, 2004). The pairwise Pearson correlations among variables for the datasets in Figure 8.5 hint about the dependencies among variables. The variables autoregressive order n was estimated using the Akaike information criterion (Kwiatkowski et al., 1992), providing an indication of the signal-own predictability. The autoregressive orders found with this criterion are summarized in Table 8.1. Stationarity of the time-series was estimated using the Kwiatkowski–Phillips–Schmidt–Shin (KPSS) test for stationarity and is summarized in Table 8.2 (Figure 8.6).

Table 8.1 Autoregressive Orders as Calculated with the Akaike Information Criterion

Var.#	*1*	*2*	*3*	*4*	*5*	*6*	*7*	*8*	*9*	*10*
Dataset 1	2	2	1	2	2	9	7	9	9	9
Dataset 2	25	1	25							

Table 8.2 Kwiatkowski–Phillips–Schmidt–Shin (KPSS) Stationarity Tests

Var.#	*Dataset 1*	*Dataset 2*
1	$p < 0.01$[a]	$p = 0.01$[b]
2	$p < 0.01$[a]	$p = 0.014$[b]
3	$p < 0.01$[a]	$p = 0.01$[b]
4	$p < 0.01$[a]	
5	$p < 0.01$[a]	
6	$p < 0.01$[a]	
7	$p = 0.04061$[b]	
8	$p = 0.005843$	
9	$p = 0.04314$[b]	
10	$p = 0.02301$[b]	

[a] A highly significant value ($p < 0.01$).
[b] Indicates a significant value ($p < 0.05$).

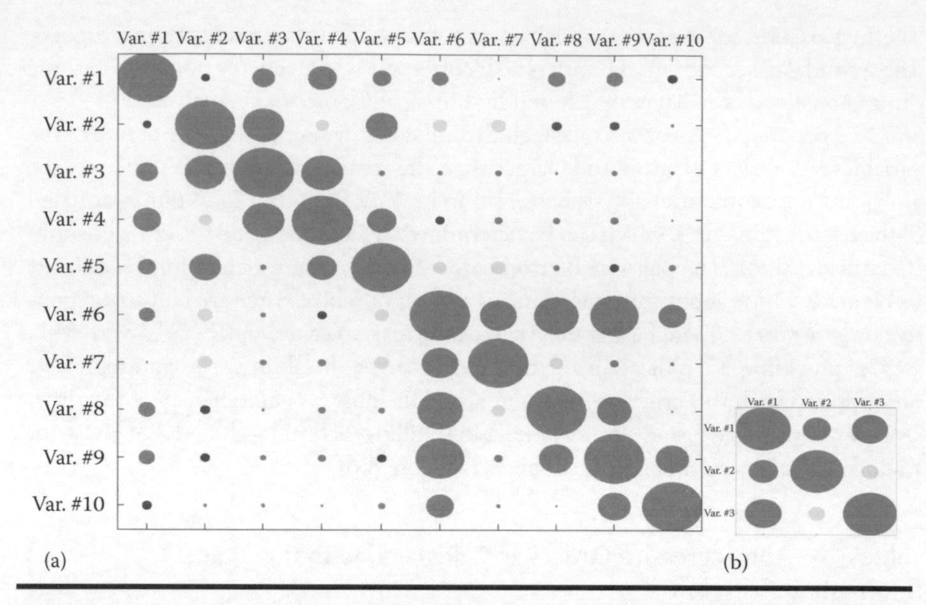

Figure 8.6 Pairwise Pearson correlations among variables for the datasets. Circle size is proportional to correlation coefficient *r*. (a) Dataset 1 and (b) Dataset 2.

Methodology for Experiments

From the datasets, specific samples were hidden to simulate missing values in three different fashions:

- *Random missing data (RMD):* Ghosted samples were chosen at random. Ghosted data accounts for 10% of each variable.
- *Random missing blocks (RMB):* Ghosted samples were chosen in blocks to have consecutive subseries of missing data. Ghosted data accounts for 10% of each variable. However, the location of the ghosted block and the number of blocks is random.
- *All missing data (AMD):* One full variable was ghosted at a time. Reconstruction can only occur from related information.

For each fashion, 10 train/test pairs were prepared for a ten-fold cross-validation. Note that the AMD has d test for each train case where d corresponds to the number of variables in the dataset. After preparation of the ghosted test datasets, reconstruction was attempted by means of the following techniques:

- *Static Bayesian network (BN):* Discretization was set to 5 equidistant intervals. Structure was learned using the PC algorithm (Spirtes et al., 2000).
- *Autoregressive Bayesian network (AR-BN):* Autoregression order was fixed to $<p, q>=<1,1>$. Vertical (intra-stage) structure was learned using the PC algorithm. Equidistant intervals were used at all times, with the number of intervals being

either 4 or 5, as bounded by memory limitations. The exemplary network for the dataset 2 is illustrated in Figure 8.5.
  - Linear interpolation (LI).
  - Cubic spline interpolation (CSI).
  - Autoregressive models (AR(1)).
  - Autoregressive models (AR(*n*)). Order *n* was chosen according to Table 8.1.

Notwithstanding, during the preparation of the train/test sets, some of the test sets did contain a number of samples lower than the autoregressive order, i.e., AR order 25 for dataset 1, variables 1 and 3. In those cases, the highest possible order was chosen based on the number of available samples.

As indicated earlier, for each reconstruction technique and ghosting fashion, a 10-fold validation was made. Since the AMD scenario can only be reconstructed from related information, this scenario cannot be resolved by interpolation or autoregressive models.

Therefore, the experiments were applied in two datasets, in three scenarios, using six techniques, and repeated 10 times. In total, 280 simulations were executed using MATLAB and Hugin (Andersen et al., 1989). For 3 simulations, mistakes in the pipeline from training to test were detected, and their results not included for further analysis. Statistical analysis was carried out in R.*

Results and Discussion

An example of the reconstruction with the different techniques is illustrated in Figure 8.7. The thick line indicates the original time series with the complete values. The rest of the plots correspond to all other techniques: static Bayesian network, AR-BN, linear interpolation, spline interpolation, AR(1), and AR(*n*). The three graphs correspond to the reconstruction of the three variables of dataset 2. The experiments correspond to the RMD scenario, that is, every element of the three time-series of dataset 2 is considered missing and is estimated with all the techniques covered in this chapter. The evaluation is made with respect to the observed element (thick line). The solid line corresponds to AR-BN technique. Qualitatively, AR-BN performs well, especially for variables 1 and 2. Linear interpolation also has a nice performance. The difference between linear interpolation and AR-BN is that the first performs well when the variable has low correlation between each other, while AR-BN takes into account the relation between all variables in a domain.

Figure 8.8 summarizes the errors incurred by each technique, according to the error metrics described in the section "Error Metrics for the Estimation of Missing Values." Bars correspond to average values and error lines indicate standard deviation.

---

* R is a language and environment for statistical computing and graphics, see http://www.r-project.org/.

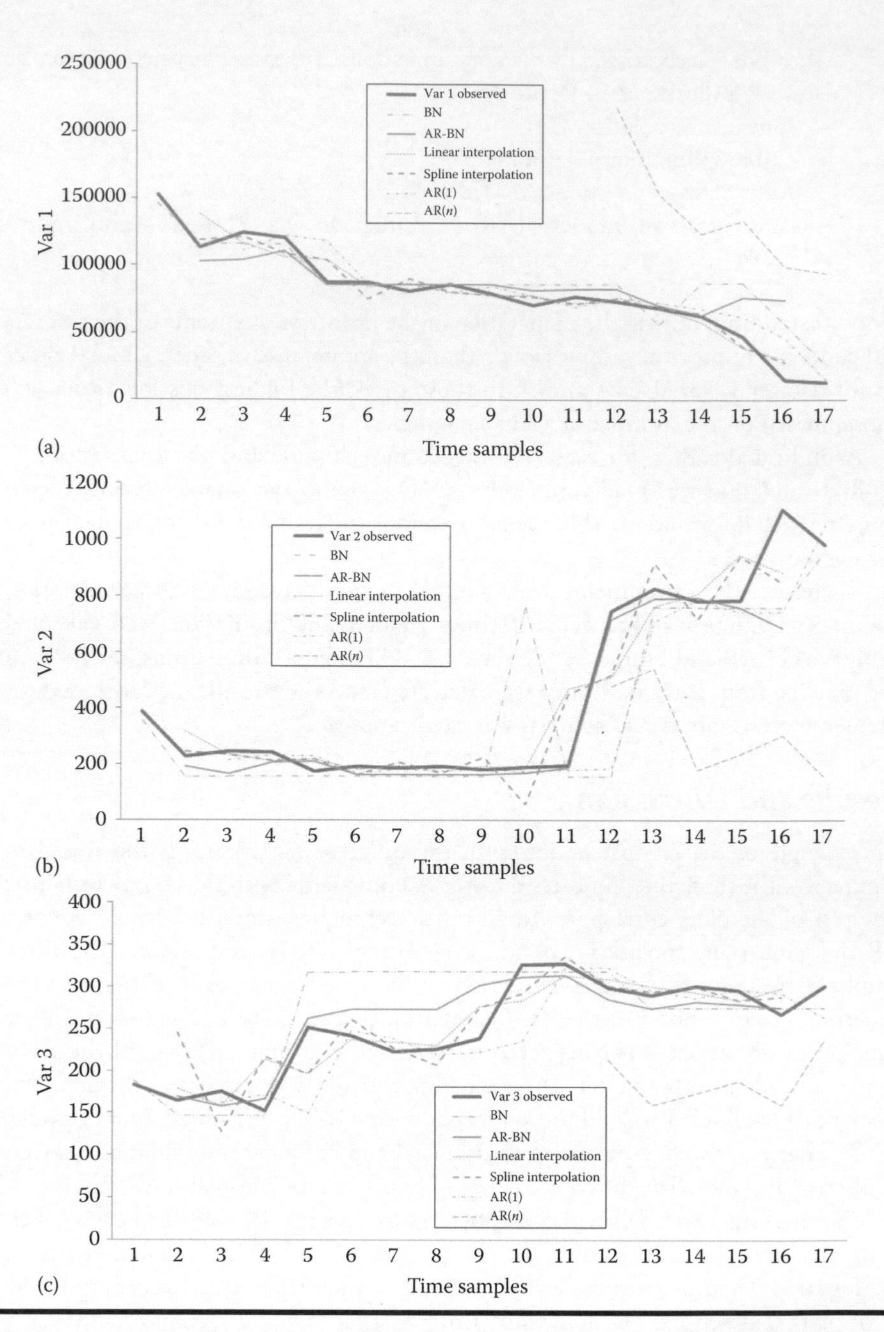

Figure 8.7 Example of the missing data estimation using the different techniques. The example corresponds to the three different variables of dataset 2, respectively, for an RMD scenario. For this example, each sample of the time series is hidden one at a time, and the missing sample is estimated using the rest of the series as necessary by the different estimation techniques.

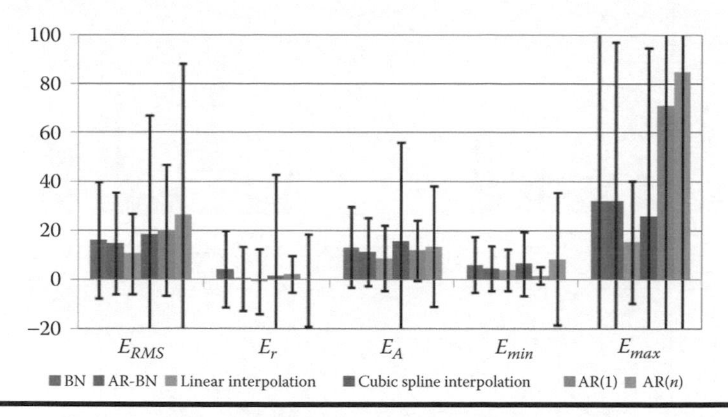

Figure 8.8 Reconstruction errors incurred by each technique across datasets, scenarios, folds, and variables. Bars and error lines correspond to average values and standard deviation, respectively.

Figure 8.9 provides a more detailed view by dataset and error metric. The left column in the figure corresponds to dataset 1, and the right column corresponds to dataset 2. The rows in the figure correspond to every error metric described in the section "Error Metrics for the Estimation of Missing Values." First, E_{rms}, next E_r, E_a, E_{min}, and E_{max}. Bars and error lines correspond to average values and standard deviation, respectively. Inside each graph, the three scenarios of missing values are exposed: random missing data (RMD), random missing blocks (RMB) and all missing data (AMD), as explained earlier.

From this detailed view, it can be appreciated that the proposed AR-BN achieves a good compromise in the reconstruction across different scenarios, datasets, and error metrics. Unexpectedly, linear interpolation achieves better overall reconstruction than the more advanced spline interpolation. Classical autoregressive models achieve reasonable performance but are highly unstable in their predictions as demonstrated by the large standard deviations coupled with disparate differences between E_{min} and E_{max}.

In order to clearly understand the meaning of these results, let us revise a portion of the information in Figure 8.9. Consider the left column, corresponding to dataset 1 formed by 10 variables. The first three rows show the measured performance of all methods with three parameters: root mean square, average percent relative, and absolute average percent relative. The literature considers that the absolute percent relative error (E_r) is an important indicator of the accuracy of the models (Osman et al., 2001). In this indicator (second row of Figure 8.9), AR-BN obtains low values for missing data and missing blocks compared with other traditional methods. However, for comparing the absolute average percent relative error (E_a), the values favored other methods.

Considering the parameters (E_min) and (E_max), the minimum values are better for methods that have no interplay between variables, but the E_max parameter

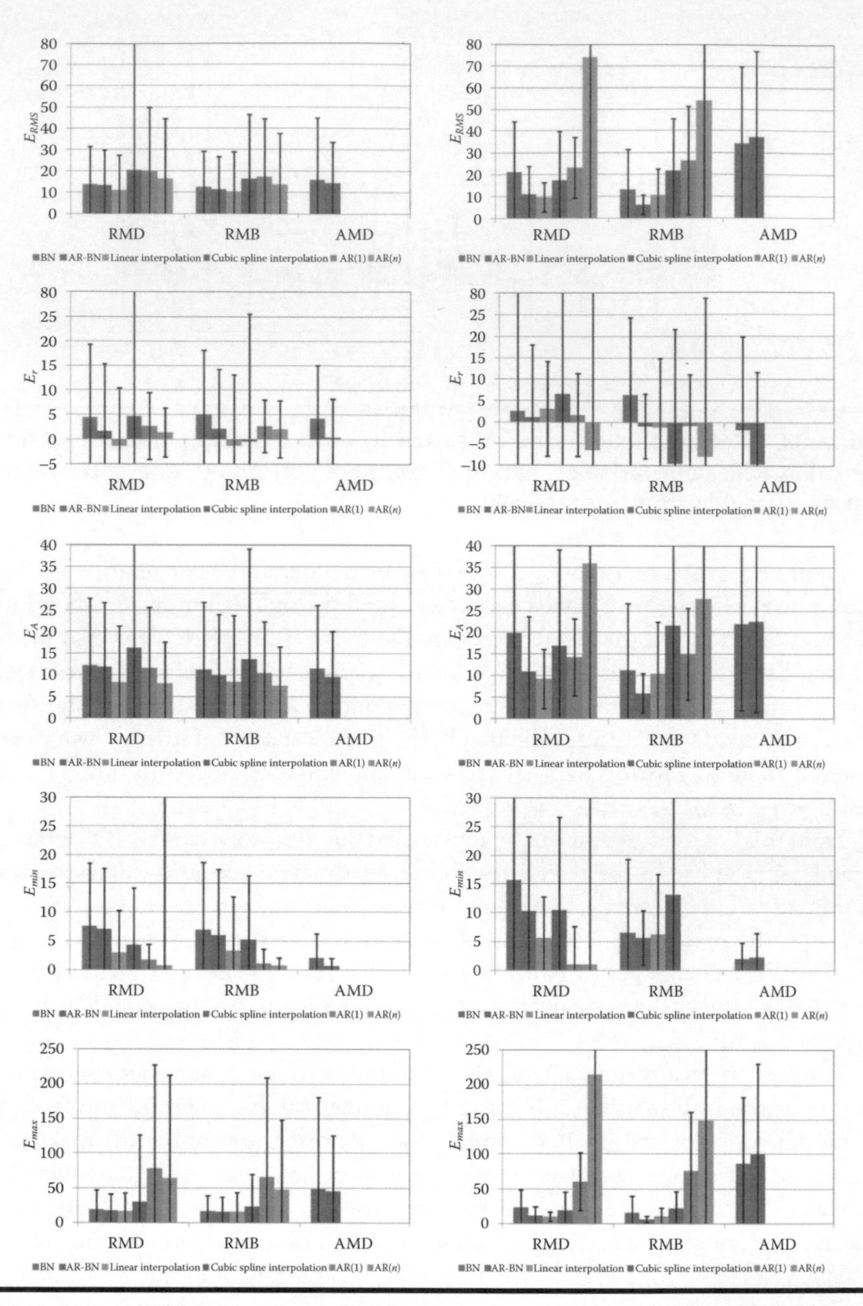

Figure 8.9 Reconstruction errors incurred by each technique across folds and variables. Columns correspond to dataset. Left: dataset 1; Right: dataset 2; Rows correspond to different error metric: from top to bottom: E_{rms}, E_r, E_a, E_{min}, and E_{max}. Bars and error lines correspond to average values and standard deviation, respectively.

favored methods that considered relations between the variables in the dataset. Notice the bottom-left graph of Figure 8.9; the performance of all these methods are similar for both missing data and missing block.

Limits of Missing Data Estimation Approaches

Figure 8.10 relates the variable feature space given by the variable autoregressive order and its average relation to all other variables in its dataset (avg_r) against the

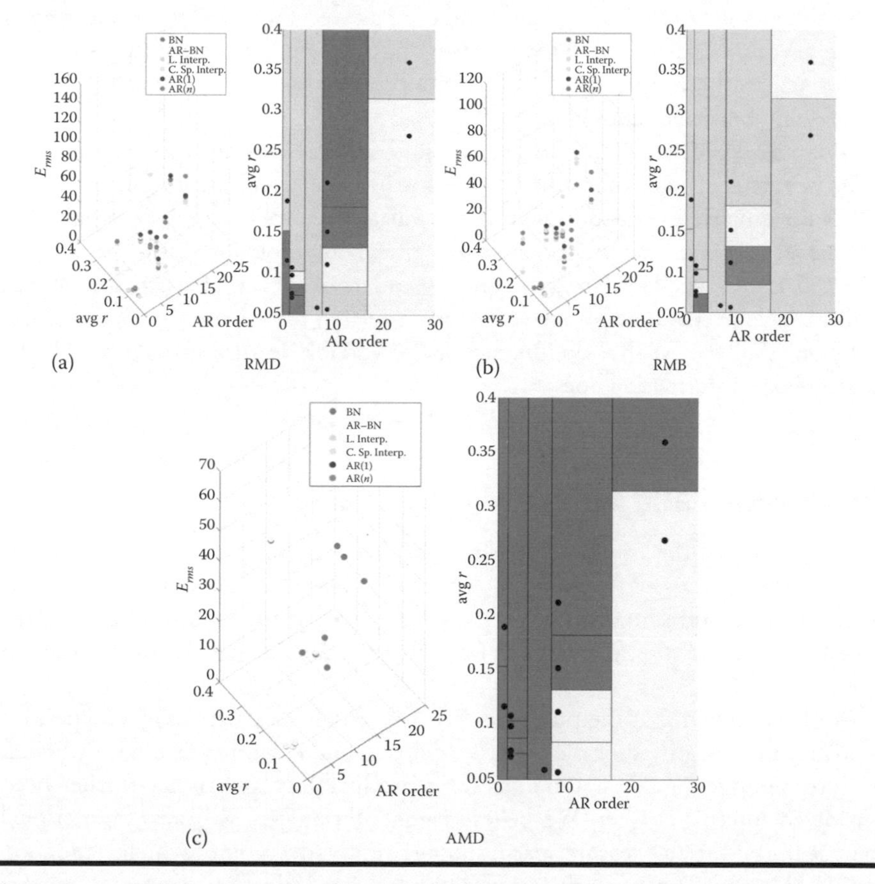

Figure 8.10 Relation between the variable feature space and the techniques for the three different scenarios: (a) random missing data (RMD), (b) random missing blocks (RMBs), and (c) all missing data (AMD). Left: Scatter plots of the variable feature space versus the error for each variable reconstructed through different techniques. The technique that achieves the lowest error is considered to dominate the region of the variable feature space. In order to determine the region of the variable feature space, the different feature vectors for each of the variables for both datasets are used as seed-vector quantizers for establishing a Voronoi partition. Each region of the Voronoi parcellation is then grayscaled according to the dominant technique.

dominant technique for the three scenarios: (1) for missing data, (2) for missing block, and (3) for missing variable. On the left side of each scenario is a scatter plot of the variable feature space versus the error for each variable reconstructed through different techniques. The technique that achieves the lowest error is considered to dominate the region of the variable feature space. The dominant technique is that which affords the lowest error in a particular region of the variable feature space. On the right side, regions are calculated using the Voronoi partition. In order to determine the region of the variable feature space, the different feature vectors for each of the variables for both datasets are used as seed-vector quantizers for establishing a Voronoi partition. 10 seed vectors appear from dataset 1 plus 3 seed vectors for dataset 2. Each region of the Voronoi parcellation is then colored according to the dominant technique.

It can be appreciated how the use of one technique over the other is subjected to the characteristics of the variable in terms of its autoregressive information, as well as the amount of dependency that the variable shares with fellow variables in the dataset as hypothesized. In particular, linear interpolation performs particularly well when the estimated autoregressive orders of the variables are low. When a full variable needs to be reconstructed from related information (scenario (c)), it is obvious that the AR-BN dominance of the variable feature space grows as the autoregressive information does.

Conclusions and Future Work

We have explored the relation between a variable feature space represented by its autoregressive order and its relation to other variables in its dataset against different reconstruction techniques. Our results suggest that the interplay between the variable's characteristics in the dataset dictates the most beneficial reconstruction option.

We have shown that the proposed AR-BN achieves a particularly competitive reconstruction regardless of the scenario, dataset, and error metric used. Although we have reported signals stationarity for reproducibility, it has not further been considered for this chapter. We believe signal stationarity will also be a critical element in the variable feature space, supporting the decision of which estimation technique to use. Consequently, we plan to explore its effect.

The AR-BN model can be trivially extended to any level of autoregression and can be easily adapted for nonnumerical data. In this sense, different autoregressive stages, whether past or future, must be added "in parallel" rather than "in series" so that these observations can be appreciated through the Markov blanket. We believe the proposed AR-BN profits from both within-variable information and statistical dependencies across variables, thus representing a valuable tool for the estimation of missing data in incomplete databases.

References

Abraham, B., and G. E. P. Box. Bayesian analysis of some outlier problems in time series. *Biometrika*, 66(2):229–236, 1979.

Akaike, H. Fitting autoregressive models for prediction. *Annals of the Institute of Statistical Mathematics*, 21(1):243–247, 1969.

Andersen, S. K., K. G. Olesen, F. V. Jensen, and F. Jensen. Hugin: A shell for building Bayesian belief universes for expert systems. In *Proceedings of the Eleventh Joint Conference on Artificial Intelligence, IJCAI*, pp. 1080–1085, Detroit, MI, August 20–25, 1989.

Balke, N. S. Detecting level shifts in time series. *Journal of Business & Economic Statistics*, 11(1):81–92, 1993.

Caruso, C. Can a social media algorithm predict a terror attack? *MIT Technology Review*, June 16, 2016.

Chatfield, C. *The Analysis of Time Series: An Introduction*. Chapman & Hall/CRC, Boca Raton, FL, 2004.

de Boor, C. *A Practical Guide to Splines*. Applied Mathematical Sciences. Springer-Verlag, New York, 1978.

Dean, T., and K. Kanazawa. A model for reasoning about persistence and causation. *Computational Intelligence*, 5(3):142–150, 1989.

Dempster, A. P., N. M. Laird, and D. B. Rubin. Maximum likelihood from incomplete data via the em algorithm. *Journal of the Royal Statistical Society. Series B (Methodological)*, 39(1):1–38, 1977.

Fukunaga, K., and D. R. Olsen. An algorithm for finding intrinsic dimensionality of data. *IEEE Transactions on Computers*, 20(2):176–183, 1971.

Hernández-Leal, P., L. E. Sucar, and J. A. González. Learning temporal nodes Bayesian networks. In *Twenty-Fourth International Florida Artificial Intelligence Research Society Conference (FLAIRS'2011)*, pp. 608–613, Palm Beach, FL, May 18–20, 2011. Association for the Advancement of Artificial Intelligence.

Herrera-Vega, J., F. Orihuela-Espina, P. H. Ibargüengoytia, E. F. Morales, and L. E. Sucar. On the use of probabilistic graphical models for data validation and selected application in the steel industry. *Engineering Applications of Artificial Intelligence*, 70:1–15, 2018.

Hoo, K. A., K. J. Tvarlapati, M. J. Piovoso, and R. Hajare. A method of robust multivariate outlier replacement. *Computers and Chemical Engineering*, 26:17–39, 2002.

Hotelling, H. Analysis of a complex of statistical variables into principal components. *Journal of Educational Psychology*, 24:417–441, 1933.

Ibargüengoytia, P. H., M. A. Delgadillo, U. A. García, and A. Reyes. Viscosity virtual sensor to control combustion in fossil fuel power plants. *Engineering Applications of Artificial Intelligence*, 29:2153–2163, 2013a.

Ibargüengoytia, P. H., U. García, F. Orihuela-Espina, J. Herrera-Vega, L. E. Sucar, E. F. Morales, and P. Hernández-Leal. On the estimation of missing data in incomplete databases: Autoregressive Bayesian networks. In *Eighth International Conference on Systems, ICONS-2014*, Sevilla, Spain, 2013b. IARIA.

Ibargüengoytia, P. H., S. Vadera, and L. E. Sucar. A probabilistic model for information and sensor validation. *The Computer Journal*, 49(1):113–126, 2006.

Koller, D., and N. Friedman. *Probabilistic Graphical Models: Principles and Techniques*. MIT Press, Cambridge, MA, 2009.

Kwiatkowski, D., P. C. B. Phillips, P. Schmidt, and Y. Shin. Testing the null hypothesis of stationarity against the alternative of a unit root. *Journal of Econometrics*, 54:159–178, 1992.

Lamrini, B., E.-K. Lakhal, M.-V. Le Lann, and L. Wehenkel. Data validation and missing data reconstruction using self-organizing map for water treatment. *Neural Computing and Applications*, 20:575–588, 2011.

Lancaster, P., and K. Salkauskas. *Curve and Surface Fitting: An Introduction*. Academic Press, London, UK, 1986.

Marr, D., and E. Hildreth. Theory of edge detection. *Proceedings of the Royal Society London B*, 207:187–217, 1980.

Mihajlovic, V., and M. Petkovic. Dynamic Bayesian networks: A state of the art. CTIT technical report series TR-CTI 36632, University of Twente. Department of Electrical Engineering, Mathematics and Computer Science (EEMCS), 2001.

Muirhead, C. R. Distinguishing outlier types in time series. *Journal of the Royal Statistical Society. Series B (Methodological)*, 48(1):39–47, 1986.

Osman, E. A., O. A. Abdel-Wahhab, and M. A. Al-Marhoun. Prediction of oil pvt properties using neural networks. In *SPE Middle East Oil Show*, 14 pp., Manama, Bahrain, March 17–20, 2001. Society of Petroleum Engineers (SPE).

Pearl, J. *Probabilistic Reasoning in Intelligent Systems: Networks of Plausible Inference*. Morgan Kaufmann, San Francisco, CA, 1988.

Peng, J., S. Peng, and Y. Hu. Partial least squares and random sample consensus in outlier detection. *Analytica Chimica Acta*, 719:24–29, 2012.

Sato, H., N. Tanaka, M. Uchida, Y. Hirabayashi, M. Kanai, T. Ashida, I. Konishi, and A. Maki. Wavelet analysis for detecting body movement artifacts in optical topography signals. *NeuroImage*, 33:580–587, 2006.

Shibata, R. Selection of the order of an autoregressive model by Akaike's information criterion. *Biometrika*, 63:117–126, 1976.

Spirtes, P., C. Glymour, and R. Sheines. *Causation, Prediction and Search*. MIT Press, Cambridge, MA, 2000.

Sucar, L. E. *Probabilistic Graphical Models: Principles and Applications*. Advances in Computer Vision and Pattern Recognition. Springer, London, UK, 2015.

Tsay, R. S. Outliers, level shifts, and variance changes in time series. *Journal of Forecasting*, 7:1–20, 1988.

Tuntun, S., M. T. Khasawneh, and J. Zhuang. New framework that uses patterns and relations to understand terrorist behaviors. *Expert Systems with Applications*, 78:358–375, 2017.

Vagin, V., and M. Fomina. Problem of knowledge discovery in noisy databases. *International Journal Machine Learning & Cyber*, 2:135–145, 2011.

Walczak, B. Outlier detection in multivariate calibration. *Chemometrics and Intelligent Laboratory Systems*, 28:259–272, 1995.

Chapter 9

Developing Cyber-Personas from Syslog Files for Insider Threat Detection: A Feasibility Study

Kevin Purcell, Sridhar Reddy Ravula, Ziyuan Huang, Mark Newman, Joshua Rykowski, and Kevin Huggins

Contents

Introduction

Since the 1990s, the U.S. Department of Defense (DoD) has pursued a doctrine of network-centric warfare. This concept grew out of the desire to harness and leverage the capabilities of networked computing systems for military advantages. These advantages are captured in the network-centric warfare tenets (Alberts, 2002):

1. Improved information sharing
2. Enhanced quality of information and shared situational awareness
3. Enabled self-synchronization
4. Increased mission effectiveness

However, there are many challenges to obtaining the full benefits from this doctrinal approach. From a technical perspective, computer networks and the communication protocols they facilitate are growing rapidly in both size and complexity. Emerging technologies similar to the Internet of Things (IoT), cloud-based computation, block-chain networks, and distributed computation networks require increasing accessibility. The combination of this increased connectivity, storage, and accessibility ushers in a suite of security concerns. For example, mobile platforms are currently capable of collecting and storing highly granular spatiotemporal data on user movements for extended periods. The storage of such information on spatial movements and information access can be dangerous if accessible to agents with nefarious intentions. Cybersecurity has made significant advances in recent years with a particular focus on detecting external threats (Chivers et al., 2009; Eberle and Holder, 2007; Zhang et al., 2008) and atypical network access (Zhang and Zulkernine, 2006). However, for all the advancements in mitigating external threats, we are still exceptionally vulnerable to internal threats.

Insider threats, or the vulnerability of networks to internal agents, has become an increasingly important concern that necessitates the study on how to identify an internal user versus an internal threat. By definition, an insider threat is someone who has legitimate access to a system, but who uses that access for a reason other than that for which the user was granted access (Bishop et al., 2009). There are currently two main ways to determine an insider threat: the multi-tiered approach (Legg et al., 2013) and the fault tree approach (Bishop et al., 2014). Both methods are reasonably robust, however, they have the complicating factor of requiring human input to analyze and make the final call if a person's actions raise to the level of an actual threat.

A persona is traditionally defined as a role or a character played by an actor. The term derives historically from the Latin term for a theatrical mask. More recently, the term persona was adopted by Carl Jung to describe the face (or mask) that a person shows to the world. In Jungian psychology personas, and an associated term, archetypes, are principle devices by which individuals express their intentions both internally and externally. This is why the term persona has become a critical tool for software development, user-centered design, and marketing. These disciplines apply the Jungian concept of personas to delineate individuals into meaningful groups based on behavioral characteristics and tendencies. We would argue that the concept of personas, or cyber-personas, could be extended to fit the cybersecurity space by offering a tool by which we can define families of network-access behavioral dynamics. An individual user may have many personas based on their current and historical network access and behavior. When they change their behavioral habits and add a new persona, they may now be moving from an insider to an inside threat. This cyber-personas definition is congruent with currently published method by Legg et al. (2013) based on the measurement tier approach.

Previous Work

There have been several approaches in recent years to modeling insider threats as a means of identifying them on computer networks before they cause significant harm to the organization. However, insider threat detection is rooted in anomaly detection. In the early 2000s, graph theoretic models were used for it. These efforts were shown to be accurate (Noble and Cook, 2003), but suffered from high false positives (Eberle and Holder, 2007). Machine learning algorithms, such as random forests, performed better than the previous graph theoretic approaches (Zhang and Zulkermine, 2006; Zhang et al., 2008).

Recent techniques that apply Bayesian analysis are similar to our approach in that they consider historical behavior, using large bodies of data in detecting insider threats (Chivers et al., 2009; 2013). However, our approach differs in that a user may have multiple personas. Moreover, in general, previous work has relied on human-in-the-loop for actually determining an insider threat (Bishop et al., 2008, 2009, 2014; Kaufman and Rousseeuw, 1990). By contrast, our personas effort is an automated, data-driven approach.

Problem Description

The objective of our analysis was to explore the potential of using unprocessed system logs to develop and monitor cyber-personas. System logs are a standard for network message logging. Each system log (syslog) message contains a wealth of information including severity labels and facility codes that indicate the message-generating software. The syslog standard is used in a variety of devices including printers, routers, and message receivers across platforms over many operating systems. This enables network engineers to easily consolidate log data from different systems in an easily

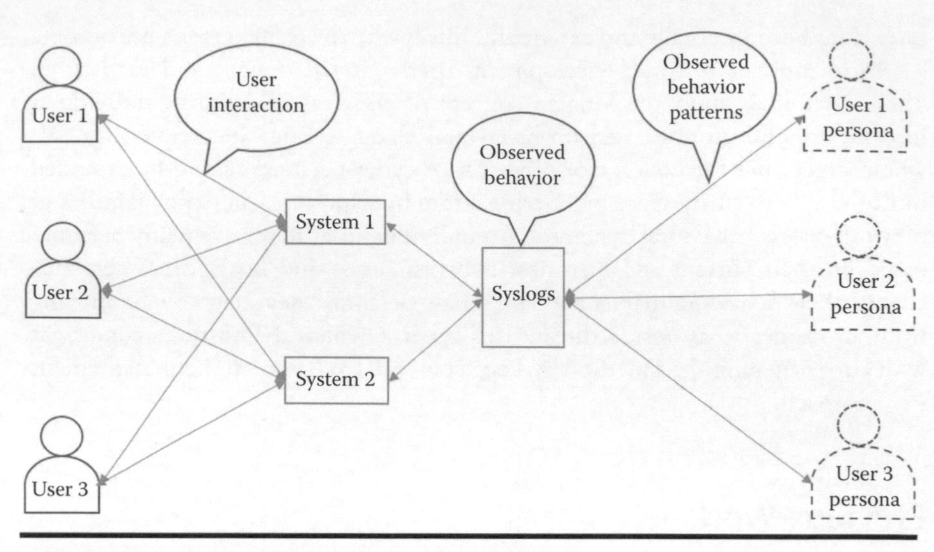

Figure 9.1 User persona based on user behavior.

accessible central repository. We propose mining a centralized syslog repository to build a system to identify user cyber-personas based on syslogs (Lonvick, 2001) and leverage those behavioral models (cyber-personas) to create a real-time system of threat detection (Figure 9.1).

Approach

To validate the concept of using data-driven personas as a system for insider threat identification and management, we envision a two-stage process that includes constructing date-driven user personas and then employing these models to identifying behavioral deviations of network users. A user's data-driven network-behavioral patterns can be viewed as their "cyber-persona." For this study, we explore the construction of user personas through an unsupervised machine learning approach applied to network syslogs obtained over a 2-month period in 2018 from an academic institution with a highly variable network user population.

To explore the feasibility of applying the data-driven persona theory to threat identification, we had to understand the extent to which syslogs could serve as a rich resource for network user behavior. Essentially, a persona defines a user or a group of users with similar network behaviors. If the variability in network behaviors is on average too high, then it becomes difficult to define a threshold by which we conclude that user behavior has changed significantly to be treated as an "anomaly" that has a high possibility of being defined as an "insider threat."

For example, a collection of five identifiable usage patterns could collectively compose a single persona. In the persona, users always login to the network both from internal source IP and external source IP during afternoons and evenings, but rarely login to the network during the daytime. Few user's destination services are blocked by the firewall. There are also kernel events occurring that are associated with the user events. The most frequent activities associated with this persona are as follows: web browsing, gaming, email, and work with other business applications, mostly during afternoons and evenings. Users that exhibit this persona might be employees working in the evening shifts. The key element that is necessary to show feasibility is a level of behavioral variability adequate to establish personas, but sufficiently limited to permit the perception of abnormal behavior.

System Architecture

The proposed system has two main components: persona construction through machine learning algorithms, and threat detection through the persona analyses (Figure 9.2). Both components consume syslogs, albeit in a different manner.

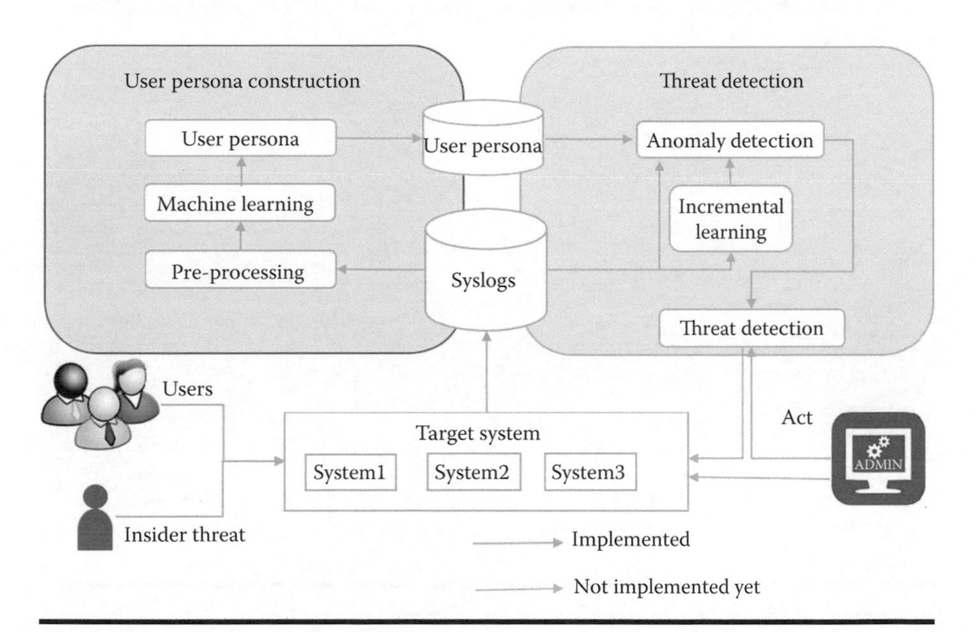

Figure 9.2 System architecture design.

When enough users exhibit an observable pattern, that pattern becomes a persona. Machine learning algorithms are used to identify and define that persona from syslog archives. When user identification is infeasible, source IPs are used as a proxy for a user.

The threat detection system compares new log entries to the corresponding user to see if that entry fits into an existing persona or not. If the existing persona is one already associated to the user, no further action is taken. If the new log entry is outside any existing personal legitimate persona, the threat system would compare it to known insider threat personas. If the level of similarity reaches a threshold, an alert is sent to the administration as a probable insider threat. If the log entry would cause the user to gain a new legitimate persona, that would generate an alert for a possible insider threat.

Implementation

We assembled a scalable system to gather and archive logs. Persona identification requires aggregation and analysis of system logs over time. The selection of flexible and robust tools for log aggregation is of utmost importance to ensure that meaningful information is captured for downstream applications. Based on the hardware environment and required flexibility, we used Logstash as the log aggregator and MongoDB as the archiving tool. We used R and Python for the analyses (Figure 9.3).

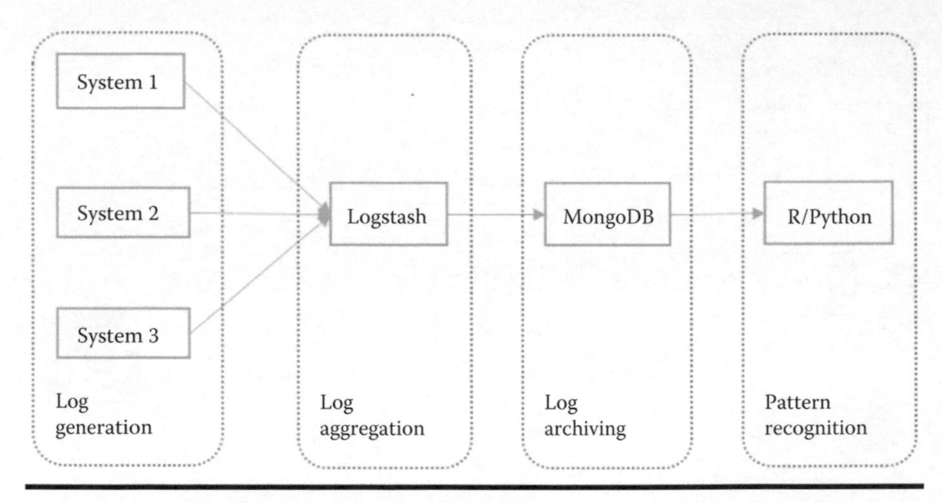

Figure 9.3 Building cyber-personas using syslogs.

Feasibility Study

To examine behavior aggregates, we applied statistical learning algorithms for clustering, for the purpose of identifying data-driven clusters of network behavior. We used random sampling to extract manageable and representative sample data (n = 10,000). The raw syslogs were parsed into individual fields before archiving in MongoDB. Further processing was done to transform timestamps and create modified vectors to capture additional information in existing fields. We used visualization to evaluate the variability in individual vectors and to discern patterns in aggregated user behavior. Cluster analysis was applied to the full panel of syslog data to analytically tackle data-driven classification in an unsupervised manner.

Most of the popular clustering techniques—k-means, hierarchical clustering, and multivariate normal distributions—are useful for analyzing continuous variables. The syslog dataset includes both continuous and categorical data vectors. In our approach to clustering data of mixed types, we applied the k-medoids algorithm (Kaufman and Rousseeuw, 1987), which is a variant of k-means. The k-medoids algorithm is similar to k-means in that it is a partitional approach but differs in how it defines the medoids, or cluster centers. While k-means has cluster centers defined by Euclidean distance (i.e., centroids), cluster centers for k-medoids are restricted to actual observations (i.e., medoids). We applied an implementation of the k-medoids algorithm in the R statistical language available in the PAM (partitioning around medoids) library (Kaufman and Rousseeuw, 1990). This preliminary study examined two approaches to syslog analysis. The first approach was event-based syslog analysis; the second was an IP-based syslog approach. The event-based method used data directly from the data source. The IP-based method used aggregated data from the data repository. For example, the count of IP address' login frequency, total data received, total data sent, the login frequency during the week, and the login frequency of hours during a day to name a few. While we observed interesting variability in the IP-level aggregated data, the need for preprocessing and lack of granularity of this dataset were significant limitations. Due to these analytical limits, we only report the outcome of our examination of the event-based data in this study.

The event-based analysis generated two clusters. Gower distance was used to calculate the distance matrix of the event syslog dataset. PAM was then used to cluster the data points. PAM's clustering procedure is based on iterative K-random entity selection. Within a medoid, each observation is assigned based on the lowest average distance. Silhouette width (Rousseeuw, 1987) identifies the similarity of observations to clusters. That is how the number of clusters was defined. The similarity is ranged from −1 to 1. The higher the similarity score, the stronger the clustering structure is (Hummel et al., 2017).

Results

We examined syslog data for the presence of cyber-personas based on the behavioral dynamics of network user activities. We utilized the Gower distance as a metric of dissimilarity between all observations in the sample dataset. We used silhouette width, a standard indicator in partitional modeling, to determine the appropriate number of clusters to be modeled (Figure 9.4). The two clusters identified in this study have a silhouette width of 0.37, which is the highest score among our clustering options. While we observed considerable differences in silhouette distance for $k = 2$ clusters relative to cluster options, a silhouette width <0.40 is traditionally considered weak structure. As this value is less than threshold value of 0.4, we maintain that the cluster structure exists, but not very strongly. Hence, further analysis is needed to validate the clusters. T-tests were used to evaluate Cluster A and Cluster B variances.

The two clusters we identified were identified as user-oriented (Cluster A) and system-oriented (Cluster B). Cluster A occupied consisted of 90% (9029) of the sample data, while cluster B was approximately 10% (971) (Figure 9.5). The two clusters showed some intriguing differences.

Cluster A's user logs can be split into multiple segments based on the rules that triggered the logs. The nature of the logs differentiated between student versus staff, login time patterns, inbound versus outbound traffic, internal process versus external process, and so on. This granularity combined with user IPs could point to repeated user behavior patterns.

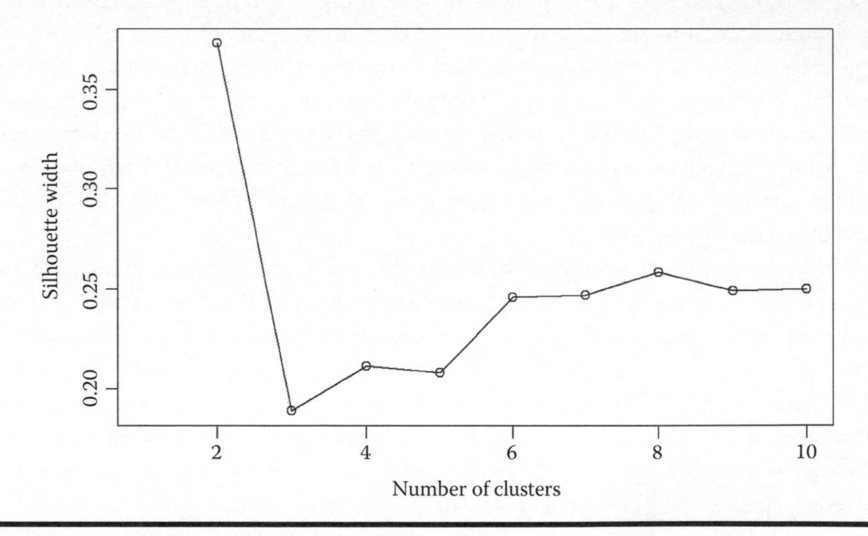

Figure 9.4 Silhouette width to identify clusters.

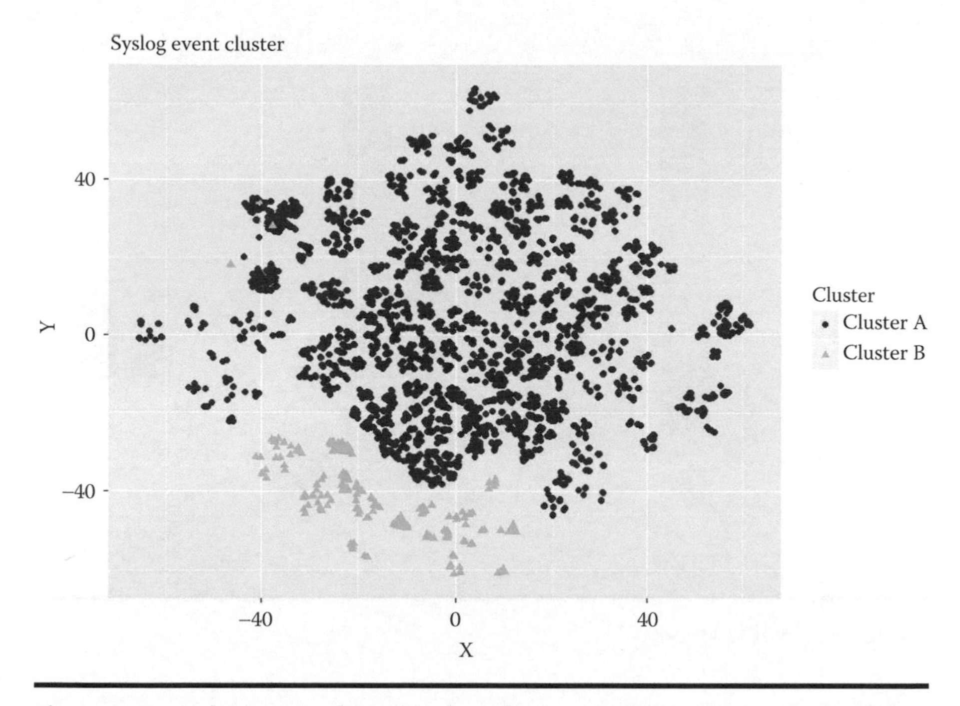

Figure 9.5 Syslog event clustering based on a random 10k sample from data source.

One of the numerous codes associated with the syslog standard is facility. This code identifies the software type generating the message syslog message. Common options include terminal, line printer, file, and remote host. Accordingly, the syslog facility code has evolved into a key data source for intrusion detection system (IDS) designers (Crosbie and Kuperman, 2001). Similarly, we believe the facility label can be useful with insider threat detection. Since it provides insights on how data is traveling through the network, facility is a critical variable that can be used to identify where syslog data originated. These linkages support the identification and analysis of personas.

Discussion

We determined that given this limited-size dataset (n = 10,000 records), there was a weak association existing between cluster and facility. Additionally, we found that data associated with Cluster A and Cluster B are significantly different, which enhanced the clustering structure with respect to syslog facility. Accordingly, we argue that syslog data is capable of providing sufficient information to define user personas. See Figure 9.6.

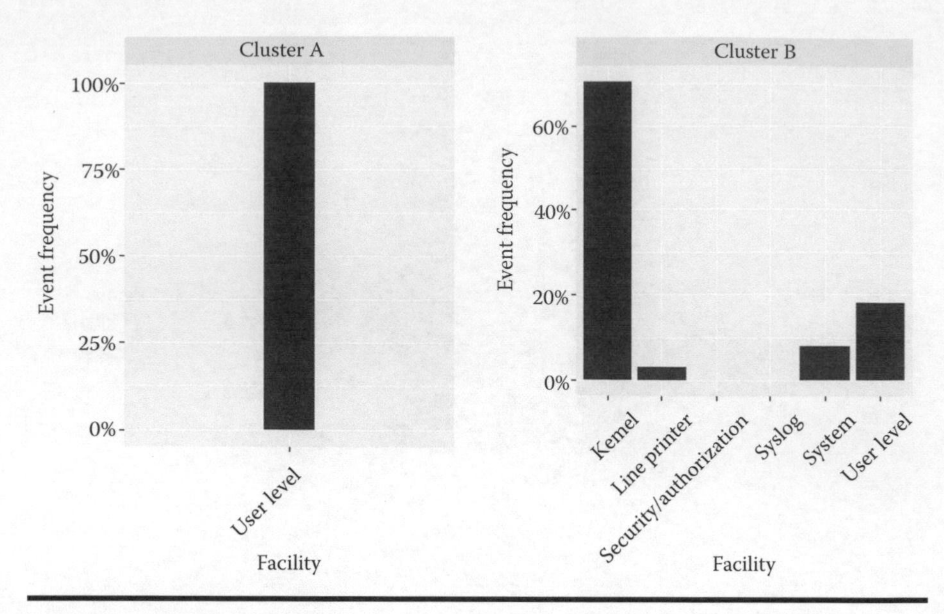

Figure 9.6 Facility syslog.

Early research (Htun and Khaing, 2012) had shown that certain syslog attributes were effective in anomaly detection for IDSs. In addition to facility codes, they identified port number (source and destination), source and destination bytes, date, time of day, and the rule attributes as an effective discriminator between clusters. Throughout our feasibility tests, we found that each of the syslog attributes mentioned were good candidates for identifying personas with common usage patterns. For example, Figure 9.7 shows the user groups based on varying interfaces accessed through different port numbers.

Data volume captured in the source and destination byte attribute was another discriminator that we found effective for clustering and identifying cyber-personas. In particular, usage patterns grouped by weekday and weekend were insightful. Figure 9.8 shows our two clusters based on network data-flow volume. Outflows were more abundant on weekends, while inbound traffic dominated during the week.

Time of day also delineated clusters via the time of day attribute in the Syslog standard. Figure 9.9 illustrates our two clusters by event per hour. Although users are active around the clock in Cluster A, the peak times were evenings. Cluster B's peak periods were early mornings.

Syslog rules were the last attribute that we considered in our feasibility study. Other researchers (Pan et al., 2015) found them relevant as well for anomaly detection for IDS's. Cluster A had syslog rules that primarily fired due to Internet and

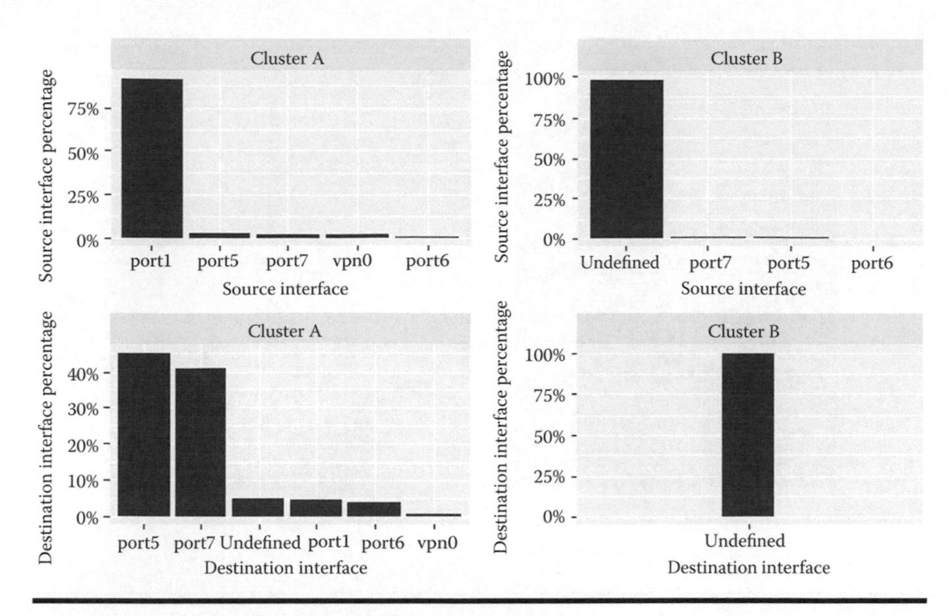

Figure 9.7 Source interface and destination interface.

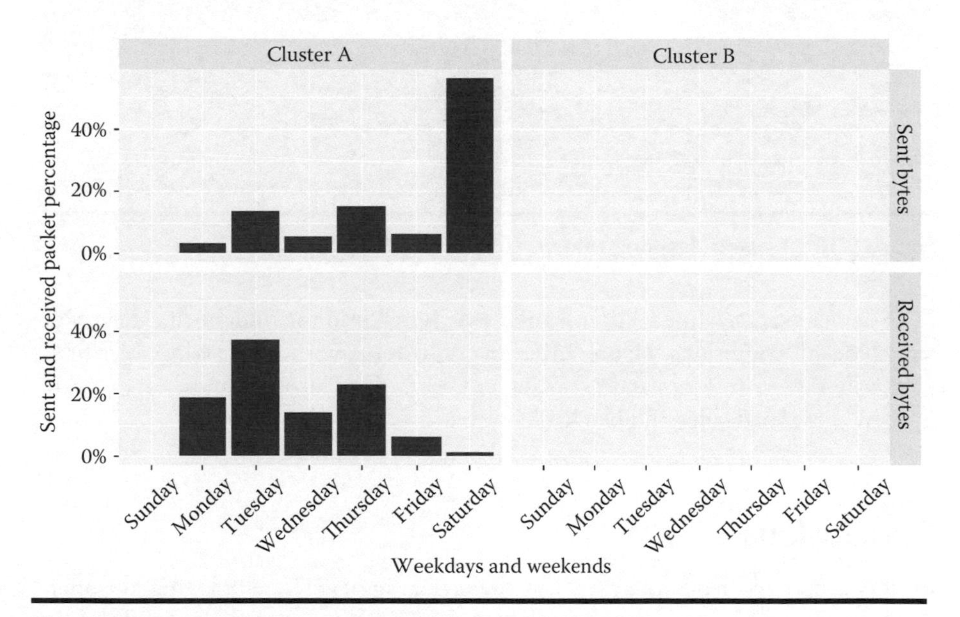

Figure 9.8 Sent and received packets.

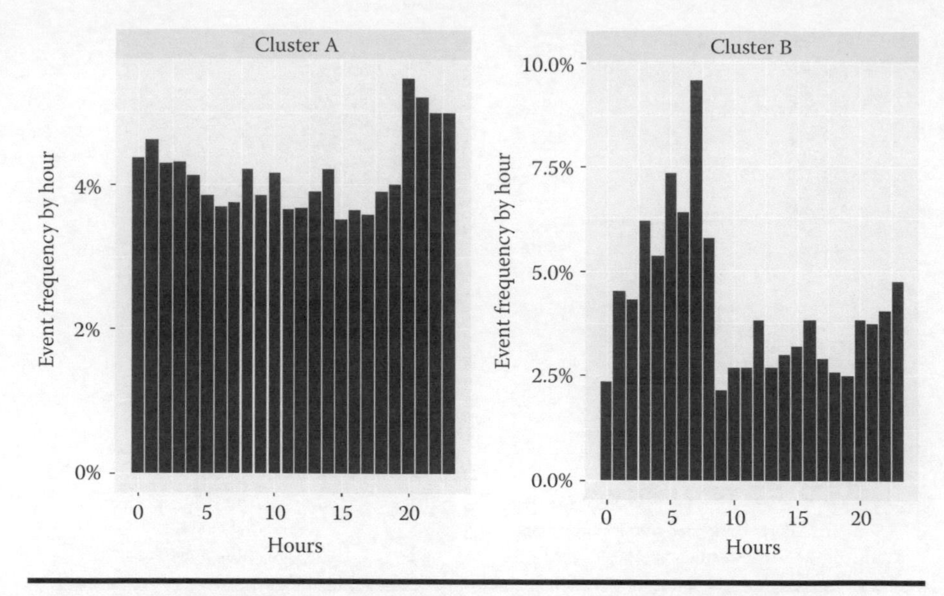

Figure 9.9 Syslog event count by hour (EST) of Cluster A and Cluster B.

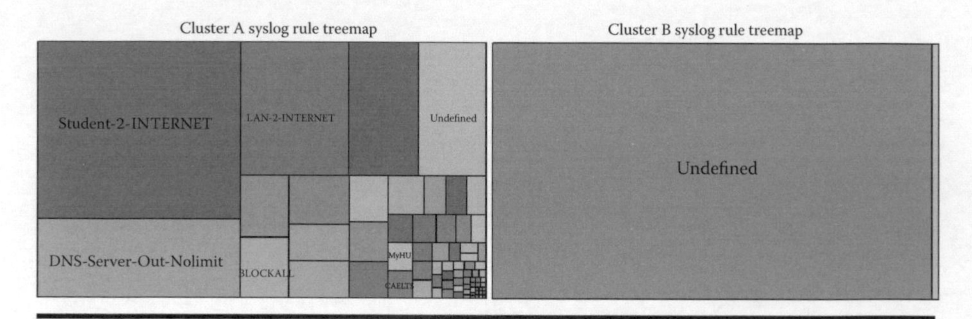

Figure 9.10 Cluster A syslog rule versus Cluster B syslog rule.

DNS server events, while Cluster B rules mostly fell into the undefined category or were blockall conditions. Figure 9.10 shows Cluster A with most syslog rules fired primarily due to Internet and DNS server events. In Cluster B, most events were fired by undefined rules and blockall conditions.

Conclusions

In this work, we have analyzed user behavior aggregates, employing unsupervised machine learning techniques on sample data from syslogs generated during user interactions with the network. Our objective was to conduct a rudimentary assessment of the ability to discern behavioral personas based on traditional network log files with little pre-processing. We found that there was structure in using fast, highly efficient, and established analytics protocols. Going forward, we plan to build both detailed user personas and a threat detection system leveraging a data-driven, automated approach.

We believe that cyber-personas, as we have identified them both practically and conceptually, are of significant value for organizations that have expansive network infrastructures with complex and evolving hierarchical security procedures, such as the U.S. DoD and its network-centric warfare doctrine. More generally, the U.S. Military has all of the same security challenges of a large distributed organization with overlapping and hierarchical security levels. However, increasingly, that network of users and information exchange is including information exchange among third parties and other entities of the military and intelligence complex. Additionally, the armed services are integrating the application of a diverse ecosystem of IoT devices and their associated data streams, all of which are monitored, controlled, and maintained via a computer network. Accordingly, the security and control of this growing network of information and devices needs a command and control structure with both the flexibility and layered security that will aid in the prevention of threats from inside while avoiding the rigidity of legacy systems that have fallen short in recent years.

Appendix A: Relevant Entries in the Syslog Database Data Dictionary

Column Name	*Column Description*
rule	Rule for the syslog event
srcIF	Source interface
srcPort	Source port number
dstIF	Destination interface
dstPort	Destination port number
timestamp	UTC time when a syslog event occurred
facility	Processes or daemons that generate syslog
sentBytes	How many bytes sent
receivedBytes	How many bytes received

References

Alberts, D. S. (2002). *Information Age Transformation: Getting to a 21st Century Military.* Washington, DC: CCRP Publications.

Bishop, M., Conboy, H. M., Phan, H., Simidchieva, B. I., Avrunin, G. S., Clarke, L. A., and Peisert, S. (2014, May). Insider threat identification by process analysis. In *Security and Privacy Workshops (SPW), 2014 IEEE* (pp. 251–264). IEEE. doi:10.1109/spw.2014.40.

Bishop, M., Engle, S., Peisert, S., Whalen, S., and Gates, C. (2008, September). We have met the enemy and he is us. In *Proceedings of the 2008 Workshop on New Security Paradigms* (pp. 1–12). ACM. https://doi.org/10.1145/1595676.1595678.

Bishop, M., Engle, S., Peisert, S., Whalen, S., and Gates, C. (2009, January). Case studies of an insider framework. In *System Sciences, 2009. HICSS'09. 42nd Hawaii International Conference on* (pp. 1–10). IEEE. doi:10.1109/hicss.2009.104.

Chivers, H., Clark, J. A., Nobles, P., Shaikh, S. A., and Chen, H. (2013). Knowing who to watch: Identifying attackers whose actions are hidden within false alarms and background noise. *Information Systems Frontiers*, 15(1), 17–34. doi:10.1007/s10796-010-9268-7.

Chivers, H., Nobles, P., Shaikh, S. A., Clark, J. A., and Chen, H. (2009, June 15–19). Accumulating evidence of insider attacks. In *The 1st International Workshop on Managing Insider Security Threats* (pp. 34–51). West Lafayette, IN: Purdue University.

Crosbie, M. J., and Kuperman, B. A. (2001). A building block approach to intrusion detection. Retrieved from http://www.cs.oberlin.edu/~kuperman/research/papers/kuperman2001raid.pdf.

Eberle, W., and Holder, L. (2007). Anomaly detection in data represented as graphs. *Intelligent Data Analysis*, 11(6), 663–689.

Htun, P. T., and Khaing, K. T. (2012). Anomaly intrusion detection system using random forests and k-nearest neighbor. *Probe*, 41102(4107), 2377.

Hummel, M., Edelmann, D., and Kopp-Schneider, A. (2017). Clustering of samples and variables with mixed-type data. *PLoS One*, 12(11), e0188274. doi:10.1371/journal.pone.0188274.

Kaufman, L., and Rousseeuw, P. J. (1987). Clustering by means of medoids. In Y. Dodge (Ed.), *Statistical Data Analysis Based on the L1L_{1}–Norm and Related Methods* (pp. 405–416). Amsterdam, the Netherlands: North-Holland.

Kaufman, L., and Rousseeuw, P. J. (1990). Partitioning around medoids (Program PAM). In *Finding Groups in Data: An Introduction to Cluster Analysis.* Hoboken, NJ: John Wiley & Sons. doi:10.1002/9780470316801.ch2.

Legg, P., Moffat, N., Nurse, J. R., Happa, J., Agrafiotis, I., Goldsmith, M., and Creese, S. (2013). Towards a conceptual model and reasoning structure for insider threat detection. *Journal of Wireless Mobile Networks, Ubiquitous Computing, and Dependable Applications*, 4(4), 20–37.

Lonvick, C. (2001). The BSD syslog protocol. doi:10.17487/rfc3164.

Noble, C. C., and Cook, D. J. (2003, August). Graph-based anomaly detection. In *Proceedings of the Ninth ACM SIGKDD International Conference on Knowledge Discovery and Data Mining* (pp. 631–636). ACM. doi:10.1145/956804.956831.

Pan, D., Liu, D., Zhou, J., and Zhang, G. (2015). Anomaly detection for satellite power subsystem with associated rules based on Kernel Principal Component Analysis. *Microelectronics Reliability*, 55, 2082–2086. doi:10.1016/j.microrel.2015.07.010.

Rousseeuw, P. J. (1987). Silhouettes: A graphical aid to the interpretation and validation of cluster analysis. *Journal of Computational and Applied Mathematics*, 20, 53–65. doi:10.1016/0377-0427(87)90125-7.

Zhang, J., and Zulkernine, M. (2006, June). Anomaly based network intrusion detection with unsupervised outlier detection. In *Communications, 2006. ICC'06. IEEE International Conference on* (Vol. 5, pp. 2388–2393). IEEE. doi:10.1109/icc.2006.255127.

Zhang, J., Zulkernine, M., and Haque, A. (2008). Random-forests-based network intrusion detection systems. *IEEE Transactions on Systems, Man, and Cybernetics, Part C (Applications and Reviews)*, 38(5), 649–659. doi:10.1109/tsmcc.2008.923876.

Chapter 10

Analytics for Military Training in Virtual Reality Environments

Miguel Pérez-Ramírez, Benjamin Eddie Zayas-Pérez, José Alberto Hernández-Aguilar, and Norma Josefina Ontiveros-Hernández

Contents

Introduction

The evolution of Virtual Reality (VR) technology has resulted in four major categories: non-immersive environments, in which users are not intended to have the sensation of being inside a virtual environment; immersive environments that create the feeling of being inside the environment; augmented reality environments that are composed of real images overlaid with virtual components, text, and 2D images, so that some aspect of the real image is enhanced (Burdea and Coiffet, 2003; Perez et al., 2004); and mixed reality (Milgram and Kishino, 1994), in which users interact with a virtual environment, but the system is aware of the user's movement and position in real space. Building virtual environments involves multiple disciplines and is influenced by other technologies, such as Big Data and analytics. For example, user training is one of the most common applications of VR technology, and we are now seeing training applications based on the integration of analytics with VR environments. This symbiosis creates more powerful applications for end users. These systems have been applied in a wide range of fields; of these, the military has become a premier user of VR technology.

In this chapter, we will review how the integration of VR and analytics is providing a new generation of applications within the military.

The remainder of this paper is organized as follows: The section "Virtual Reality Fundamentals" explains the main concepts of virtual reality. The section "Big Data and Big Data Sources" in turn reviews some of the main Big Data concepts. The section "Analytics" reviews the concepts of analytics. The section "Virtual Reality and Analytics" briefly reviews some examples of how VR and analytics have been integrated in a new generation of applications. The section "Virtualytics in the Military," discusses how VR and analytics can be applied within military domain and describes an architecture. Finally, the section "Conclusions" presents our conclusions followed by a list of references.

Virtual Reality Fundamentals

There are a variety of terms and perspectives associated with the concept of virtual reality. Jaron Lanier coined the term "virtual reality" in 1989, defining it as a "3D interactive environment generated by a computer in which a person is immersed" (Machover and Tice, 1994). "Virtual environments" consist of an interactive deployment of images enhanced by non-visual components such as audio and tactile feedback to convince users of being immersed in a synthetic space (Ellis, 1994). For the purposes of this paper, we prefer the following definition:

> Virtual Reality is the electronic representation, partial or complete, of a real or fictitious environment. This representation can include 3D graphics and images, has the property of being interactive, and might or might not be immersive. (Perez et al., 2004)

It is worth mentioning that unlike Lanier's definition, immersion is not mandatory to make the claim that a system is based on virtual reality. In fact, there is a range of degrees of immersion, from non-immersive VR as shown in Figure 10.1a, to fully immersive VR as shown in Figure 10.1b. In the former, a user can interact with the VR system using traditional interfaces such as a mouse and keyboard, while in the latter, a variety of interfaces may be used to stimulate user senses to provide him/her the feeling of being within a virtual environment and to detect user actions within such an environment. Between these extremes, we find augmented and mixed reality systems. Figure 10.1c depicts an augmented reality (AR) display that superimposes virtual images over real images (Logie Lab, 2016). Some authors (Höllerer and Feiner, 2004) also consider that text superimposed to real images can be included as AR; this was provided previously as computer-generated annotations deployed in HUD (Head up Displays). They also include the entertainment industry within the scope of AR and mention the Terminator annotated vision (Cameron, 1984). Now we have the mobile version (Figure 10.1c) and devices such as glasses. AR might provide users not only with textual annotation but also with 3D graphics. Many application fields benefit from AR, such as medicine (Fuchs et al., 1998) and even the field of archaeological heritage and museum exhibitions (Wojciechowski et al., 2004). The popular *Pokémon Go* is also a well-known AR application. Microsoft's Kinect system

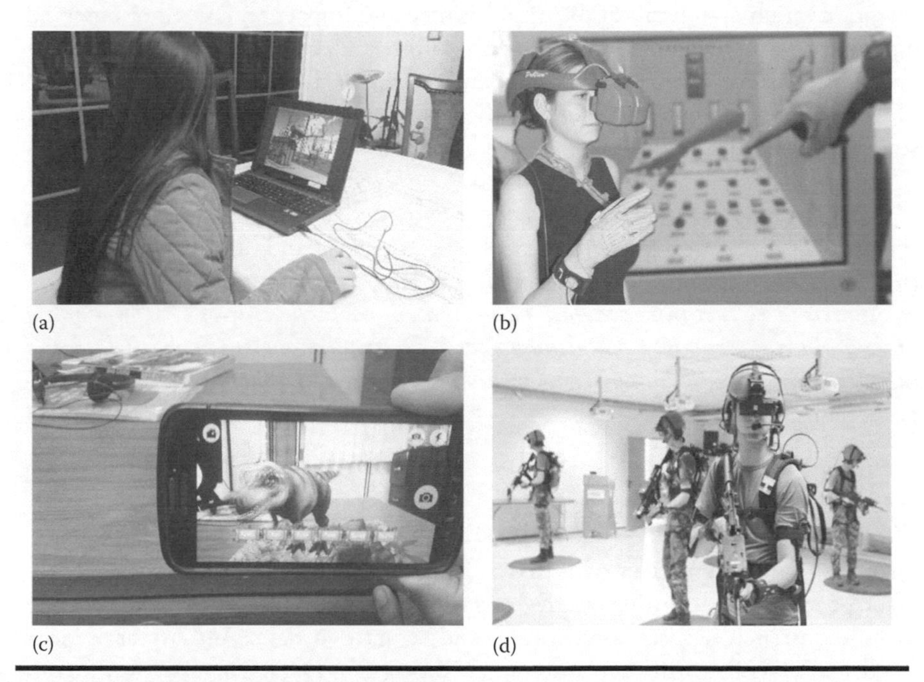

Figure 10.1 Types of virtual reality: (a) non-immersive VR; (b) immersive VR; (c) augmented reality; and (d) mixed reality.

is an example of a device that allows for developing mixed reality applications (Figure 10.1d), where a user's movements in real space are materialized inside of a non-immersive virtual environment (Parkin, 2015).

The specifics of the field of application of VR technology dictate what level of immersion is appropriate. In addition, because non-immersive VR systems do not require specialized peripheral devices to allow the user to interact with the environment, the associated costs of non-immersive systems are generally lower.

Big Data and Big Data Sources

Big Data

In a report from IBM, the authors define Big Data (BD) in terms of its main features, also called dimensions, which describe this concept in terms of the so called "classic V's" (Schroeck et al., 2012), namely: *Volume*, *Variety*, *Velocity*, and *Veracity*. Some authors add other V's to emphasize the relevance of other BD features (Biehn, 2013). One is *Viability*, which suggests the importance of carefully selecting attributes and factors that are most likely to predict outcomes that matter most to businesses. In fact, the author proposes using statistics to either discriminate or integrate variables in terms of their significance and impact on desired or observed outcomes within a specific hypothesis test. Another *V* is for *Value*; dealing with big amounts of information demands companies to create a supporting infrastructure. However, if this is not cost-effective, then companies might just run the risk of simply creating big costs without creating the value that translates into a business advantage.

Olshannikova et al. (2015) provide a extended version of Gartner's definition, which integrates and summarizes all traditional and other additional V's: *Big Data is a technology to process high-volume, high-velocity, high-variety data or datasets to extract intended data value and ensure high veracity of original data and obtained information that demand cost-effective, innovative forms of data and information processing (analytics) for enhanced insight, decision making, and processes control.*

Big Data Sources

There is a huge amount of data that is generated in the Internet every second. Data generation is growing yearly by various sources and types of data; these include unstructured, semi-structured, and structured data. Authors have identified and classified different sources generating Big Data. For instance, Soares (2012) distinguishes five sources of Big Data, including: (1) web and social media,

(2) machine-to-machine data, (3) big transaction data, (4) biometrics, and (5) human-generated data. Whereas within the military, Haridas (2015) mentions three types of sources:

1. *Machine-generated data*: Here he includes movement of different vehicles such as ships and aircrafts, drones, sensors, satellites, Unmanned Aerial Vehicles (UAVs), reconnaissance aircraft, sensors, and Battle Field Surveillance Radars (BFSR), among others. According to this author, the U.S. Argus ground surveillance system collects more than 40GB of information per second. It is an autonomous Real-Time Ground Ubiquitous Surveillance Imaging System and has a 1.8 gigapixel video camera with 12 frames per second (fps) and 368 sensors. It collects 6000 terabytes of data/imagery per day, which it feeds to Homeland Security.
2. *Human-generated data*: This includes biosensors attached to soldiers, and social media such as Facebook, Twitter, and YouTube, among others, that must be considered for strategy purpose.
3. *Business- and third-party-generated data*: This includes e-commerce, demographic, financial, and meteorological data.

There might be a large number of Big Data sources, but this term can be generalized by introducing the following concept.

Big Data Entity (BDE) is a business, company, event, object, device, person, group, and so on, whose datification process generates huge amounts of data that can be managed with Big Data technology and can be analyzed to extract some value by using techniques and methods of analytics.

Datification provides—if not yet the means in all cases, at least the idea—that it is possible to continuously gather data over time and manage enormous amounts of information that can then be used for some purpose. A clearer view is provided by Elliott (2013): "Datification is about taking a process or activity that was previously invisible and turning it into data." For instance, we might want to datify all physiological vital signs of a person (i.e., breathing, blood pressure, heart rate, temperature, etc.) over the course of a whole day or for many days, as is exemplified in the film *The Circle* (Ponsoldt, 2017), where all biometrics signals of Mae Holland (Emma Watson) and her parents are taken and analyzed by the Company she works for.

Going back to the concept of BDE, it can be added that they are dynamic in that the needs of the entities change over time, and also that the process of datification is constantly updating the entity's data repositories. These changes may be triggered by specific events, or they may be based on predefined periodical monitoring of data sources. The data repositories might contain the whole changes history of one BDE or of a group of interacting BDEs.

Thus, the authors above are mentioning some instances and classifications of BDE, but the concept of BDE together with the concept of datification is a tool to identify new BDEs where data can be analyzed in order to contribute some value to people or companies.

Analytics

Only the relevant concepts of analytics are addressed in this section, but just enough to allow exploration of how VR and analytics could be integrated to provide useful applications within the military.

Analytics is semantically described differently across fields but is composed generally of similar functions and methodologies. Fakete (2013) describes analytics as the science of drawing conclusions from data analysis. By making Big Data explicit, Galetto (2017) describes Big Data Analytics (BDA) as the process of examining large datasets to uncover hidden patterns, unknown correlations, market trends, customer preferences, and other useful information, and again, all these are used to make informed decisions. For Schroeck et al. (2012), analytics is the use of Big Data in the real world.

Thus, analytics refers to the process for systematic analysis of data using a variety of techniques to get insights from a large set of heterogeneous data. These techniques are based on a combination of business rules, algorithms, machine learning, data mining, statistical analysis, natural language processing, text analytics, artificial intelligence, visualization, and so on. The analytical techniques are applied to different datasets including historical and transactional data and real-time data feeds (Russom, 2011). Analytics are designed to reveal insights that facilitate the understanding, prediction, and exposition of hidden information to improve operational and business efficiency and to deliver real-world situational awareness. Analytics can be used to learn from historical behaviors and understand what happened or what is happening; to estimate the likelihood of future outcomes such as events and behaviors; and to advise on possible outcomes and a plan of actions. When analytics are applied appropriately to a specific domain, they can be used to increase efficiency, productivity, quality, reliability, and suitability, among others. In the military realm, analytics can be used for (Çintiriz et al., 2015):

- Intelligence development (intelligence, reconnaissance, and surveillance)
- Knowledge management
- Common operational picture
- Military decision-making process
- Cyber-defense information system management
- Military forensics
- Geographical data systems

Some of these are described in the following sections.

Types of Analytics

Analytic types generally fall into one of three categories: descriptive, predictive, and prescriptive. Most authors agree that descriptive analytics describe what has happened as indicated by the data. Wu (2013) called it "the simplest class of analytics." However, descriptive models identify different relationships and categories of data, summaries, and statistics (mean, variance, max, min, percentile, etc.) of single variables or group of variables. Descriptive analytics is the base for further analytical work, predictive or prescriptive analysis. This is the reason why it comprises 80% of data analytical work (Bertolucci, 2013). The most common example is simple counters, such as counts of posts, mentions, fans, followers, page views, kudos, check-ins, pins, and so on, in social media analysis.

As the name implies, predictive analytic methods examine current and historical data as well as trends and apply models in an attempt to forecast the future. These methods commonly use probabilistic models supported by historical data, which can result in predictions with a high confidence level. Some common examples of predictive analytics in business include predicting year-end sales, predicting what combinations of items a customer may purchase, and forecasting inventory levels based on a myriad of variables (Halo, 2016). Another example was dramatized in the film *Moneyball* (Miller, 2014), which is an adaptation of the book *The Art of Winning an Unfair Game* (Lewis, 2004). This story demonstrates using statistics to support the decisions to create a baseball team. The team managers had limited economic resources, so they chose players who were unwanted by other teams based on predictive statistical analysis of performance. A more ordinary example can be found in online marketing websites, such as Amazon.com, where suggestions for book purchases are offered to customers based on analysis of that customer's previous purchases with similar interests.

Prescriptive analytics also involve looking into the future, but in this case, the focus is on discovering a course of action that will result in a desired outcome (Halo, 2016). The goal is to produce information about the most high-value actions for taking preventative measures (Stimmel, 2015). Prescription requires understanding possible courses of action and the consequences of those actions to either achieve a particular state or prevent an undesirable one. This resembles the prescription issued by a physician to cure or prevent some disease. This can be also a powerful decision support tool to guide a company to a desired outcome.

Virtual Reality and Analytics

The joining of VR and analytics is a young trend known as VR Analytics (VRA). There are even endeavors that try to merge Artificial Intelligence, Big Data, and Virtual/Augmented Reality, as is the case of the company *Virtualitics*, which at the same time proposes this merged term making reference to both VR and Analytics (Patterson, 2017).

According to Osarek et al. (2016), VRA comes in two types: visualization of analytics data in VR (Figure 10.2), and analysis of VR experiences in a data warehouse.

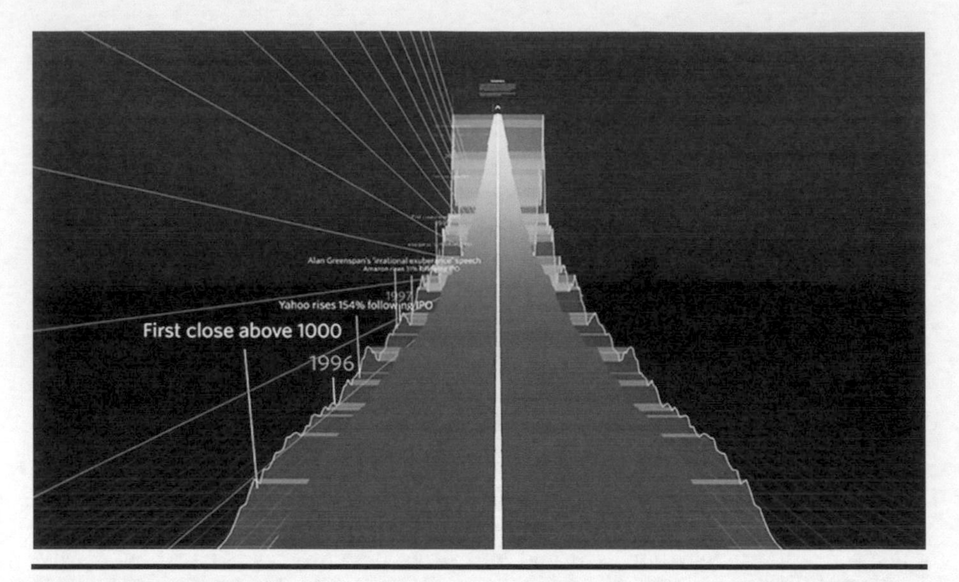

Figure 10.2 Wall Street Journal stock market data visualization journey. (Underwood, 2016.)

Regarding the VR experience, there are companies that offer analytics tools for VR systems that provide insight into how users experience VR content (Aldin, 2016; Retinad, 2016). Here, user behavior is analyzed through interactive recordings, identifying usage patterns, and trends. These systems attempt to learn what engages users and test different virtual environment configurations to build an optimal VR experience. This information can be used for various purposes; the one most frequently mentioned is the monitoring of users' preferences for commercial purposes.

On the other hand, there are also reports of visualization of analytics data in VR, for instance, iViz, an immersive, multidimensional visualization platform developed at Caltech (Underwood, 2016).

Complexity in Big Data visual analytics brings new challenges but also new tools. Analytics might allow us to manage and analyze biometrical data, for instance, to make better diagnoses and provide more effective treatment prescriptions. However, how can this information be represented within a 3D environment? This requires careful design so that the 3D representation of data makes sense; that is, it is interpretable and useful. Then it might be the case that it provides some extra value that we cannot now even imagine, providing a new view of the data and overriding some cognitive and perception limitations, resulting from data that usually is not even gathered together, much less presented in 3D. Thus, in this new field that tries to deploy a 3D representation to visualize the outcomes yielded by analytics, the challenge would be that such representations must be interpretable, and whatever the visualizations are, they must contribute *Value* to users as either individuals or organizations.

VRA can be used in the civilian world as a visualization tool for large datasets (Järvinen, 2013). For instance, VRA can be used to create a virtual underground environment from data gathered from wells and perforations or other techniques such as seismic reflection in order to aid in explorations for oil or geothermal reservoirs. We discuss military applications of VRA merging analytics and VR, approaching this as Virtualytics, following the virtualitics by Patterson (2017).

Virtualytics in the Military

VR in the Military

The various military areas—Army, Navy, and Air Force—were early adopters of virtual reality technology and remain one of the largest users (Baumann, 2010; Lele, 2013). VR technologies appear in many areas, including training, exercise support, combat simulation, weapons training, and military history education (Craig et al., 2009; Krevelen, 2007; Virtual Reality Society, 2016). More recently, VR applications have been used in the treatment of Post-Traumatic Stress Disorder (PTSD) (Rizzo et al., 2009, 2011). Training applications are the large majority of current VR implementations (Fletcher, 2013). Military training often involves learning dangerous or even life-threatening tasks, and virtual reality allows those tasks to be learned without putting the user in harm's way. That is one of the most important benefits of VR technology in the military arena, namely because it allows interacting, navigating, experiencing, and learning within virtual representations of highly dangerous environments in a safe way, free of the represented risk.

Another important benefit to the military is that the use of VR technology allows some kinds of training to be performed in a much more cost-effective way. Virtual reality flight simulators, for example, provide the opportunity for pilots to practice flight techniques and combat simulations without incurring the cost of fuel and equipment maintenance and while reducing the risk of injury and damages to expensive equipment (Craig et al., 2009). The same concept applies to ships, tanks, and other military systems as well. VR is even used for dismounted combat training so that soldiers, marines, and security personnel can face real combat situations (Haar, 2005).

There are other proposed applications of VR technologies to military training. For example, most modern military equipment includes onboard computer systems that could integrate VR technology to allow training to be conducted on the equipment or vehicle, even when the equipment and its operators are deployed in theater or at sea (Knerr et al., 2002). Collaborative VR systems would allow team training and collective skills that military operations require (Salas et al., 2002), for instance, in mission rehearsal.

Augmented Reality applications have been used by the military for some time. One common example is the helmet mounted sight (HMS) system used in combat aircraft. These systems provide basic navigation and flight information in the pilot's helmet view. Information about targets in the environment can sometimes also be provided. In some applications, the Augmented Reality information can be coupled with weapons systems; for example, the chin turret in a combat helicopter is slaved to sensors in the pilot's helmet, so the pilot can aim the chin turret simply by looking at the target (Azuma, 1997).

While many of these military examples have limited use in the civilian world, there are also applications that are useful in the civilian arena. One such application is in medical education (Dunne and McDonald, 2010; Schmitt et al., 2012). Training is conducted in a realistic 3D environment, avoiding trial-and-error situations involving real patients. This prevents the harm that could be caused by incorrect diagnosis or surgical mishaps.

In 2008, Wilson observed that VR technologies were widely useful in military applications and that these techniques could also be applied to counter-terrorism efforts. The U.S. Congress has concluded that neglecting to consider wider uses of VR technology could have a negative impact on national security (Wilson, 2008).

Analytics Applications in the Military

It is foreseen that Big Data Analytics can be useful in different ways in the military field. For instance, Haridas (2015) argues that Big Data Analytics has a significant role in the predictive capability to anticipate specific incidents. As in other fields, correct prediction allows anticipating courses of action so that facts can be prevented, for instance, as in the case of terrorism attacks, and facilitated in case of anticipated action against possible attacks. Analytics is even considered the key to modern readiness (Gittings, 2017).

However, other authors are more cautious. For instance, Pomerleau (2016) points out that before celebrating the potential benefits of Big Data Analytics in the military, a cost-benefit analysis must be performed in order to evaluate the effectiveness that analytics currently produces. This author (emphasizing one of the V's of BD) suggests making sure that data comes from consistent, reliable sources and to ensure the quality of the data (*Veracity*). He adds that dataentry errors substantially impair the best algorithms, originating wrong analysis that in turn will derive in wrong decision making.

VR and Analytics in the Military

Now, we have three different domains that we want to integrate, namely, VR, analytics, and the military. Specifically, we want to know how VR and analytics can be applied (mapped) into the military domain. In order to do so, we can approach these three domains as three dimensions in a cube where the dimension of VR and the dimension of analytics deploy their main features and benefits, then these two combined dimensions (*Virtualytics*) are mapped into different areas of the military dimension in terms of possible applications. Thus, we might define all these dimension as follows, using not all but only some main features on each dimension with illustrative purposes.

VR dimension (*VRD*): Provides features such as visualization, interactivity, and navigation. Augmented Reality can enrich real images with textual information (such as HUDs) and 3D information. Mixed reality allows interaction with real and virtual objects.

Analytics dimension (*AD*): Provides the benefits of description, prediction, and prescription.

Military dimension (*MD*): Military organization includes at least three main areas, namely, Air Force, Navy, and Army, so that the main navigation means are covered (Wikipedia, 2017b). Each of these military areas is a potential field of *Virtualytics* application. However, identification of relevant activities and processes in the MD can be helpful to distinguish more accurately possible application areas. Some examples are:

- Monitoring. This includes biometrics, *mechametrics*, social networks, geolocation (troops, vehicles, or both), etc. Here, *mechametrics* can be defined, analogously to biometrics, to make reference to the measurement of the general mechanical features of vehicles including fuel, oil, and other levels; vehicles can be trucks, tanks, planes, ships, etc.
- Training. It includes on-foot fighting, aircraft flying, aircraft fighting, ship fighting, battle rehearsal, etc.
- Decision-making process. The military decision-making process (MDMP) is a single, established, and proven analytical process (Tradoc, 1997) that helps the commander and his staff examine a battlefield situation and reach logical decisions.
- Readiness. According to Joseph Anderson, Lieutenant General, U.S. Army, "readiness is the capability of our forces to conduct a full range of military operations to defeat all enemies, regardless of the threats that they impose. It is generated through manning, training and equipping our units and leader development" (Rumbaugh, 2017).
- ISTAR (Intelligence, Surveillance, Target Acquisition, and Reconnaissance). This practice links several battlefield functions together to assist a combat force in employing its sensors and managing the information they gather (wikipedia, 2017a).
- Military operations. A set of military actions intended to accomplish a task or mission (Scott, 2011).

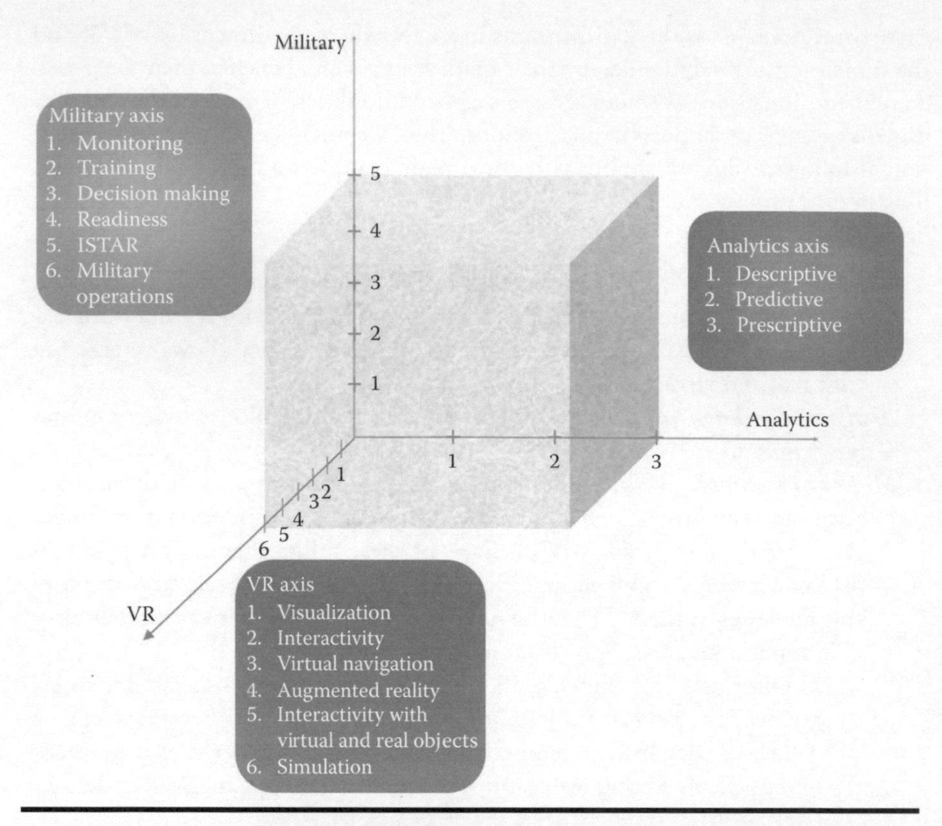

Figure 10.3 VR and analytics dimension mapped into the military dimension.

In an attempt to integrate these three dimensions, we use a cube, where features and benefits of VR and analytics axes are mapped into possible applications in the military axis (Figure 10.3). Some examples are described as follows:

- *(Descriptive analytics, Visualization)*: This combination can be applied to different purposes such as in monitoring different features including biometrics and mechametrics together with the geolocation of persons and vehicles. It might be useful to know the physical and mechanical conditions of soldiers and vehicles (planes, tanks, ships), respectively, and visualize their geographic location. Monitoring social networks might be helpful to know about opinions, tendencies, or criminal suspects and visualize their geographical location. Free 3D navigation might allow greater location accuracy and eliminate problems of losing troops and vehicles. Bio/mechametrics also would provide more insight about wounded soldiers, casualties, and damaged vehicles, and visual geolocation information might be used for rescue planning. All monitoring information would also be useful for determining military moving capacity and for decision-making processes such as planning; for instance, automatic maintenance planning, mission assignation of aircrafts, ships, and tanks, and in general, military vehicles based on actual location, intended destination, and mechametrics.

Information, including GPS coordinates, gathered in surveillance missions might be deployed in a geographical location within a 3D environment, either as mixed or augmented reality (Figure 10.4). Then the visual information could be interactively available for any planning or decision-making process (Forces, 2016).

- *(Prescriptive analytics, Interaction)*: Again, different applications can be identified. For instance, training can be prescribed for different purposes. VR is able to provide intelligent, interactive environments where nearly all aspects of military training can be covered: hand-to-hand combat, aircraft battle, ship and submarine battle, battlefield tactics, etc. These, together with biometrics, would help monitor levels of stress, mastering of a subject, correctness of decision making, aptitude, attitude, etc. The intelligence of ISTAR might use sensed information for planning its military operations and prescribe specific courses of action, which can be even rehearsed in an interactive virtual environment. See for instance, McGregor et al. (2017), for a platform that, in real time, acquires data from a first-person shooter military combat simulation game where a multisensory garment is used to collect biometrics data.
- *(Predictive analytics, Visualization)*: Predictive analytics can have a wide range of applications in preventing unwanted outcomes. Different examples can be mentioned; for instance, training injuries can be predicted from monitoring biometric data during training, and then these can be prevented. It has been already mentioned that VR allows training for dangerous activities under a

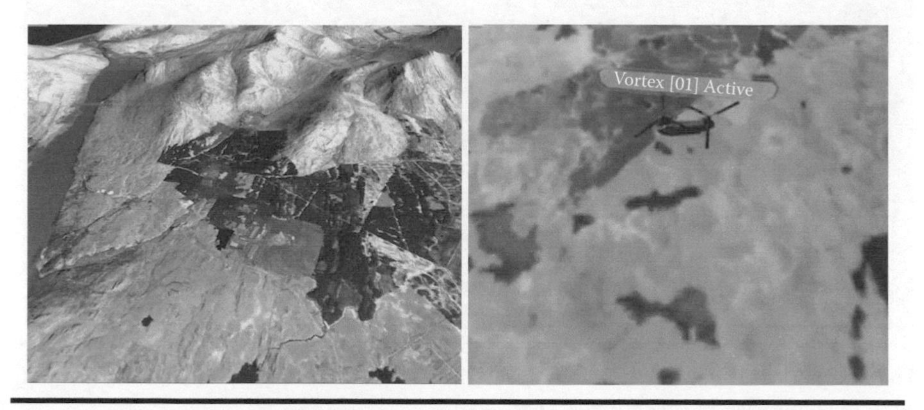

Figure 10.4 Scan of a geographical location in a 3D environment. (Forces, 2016.)

safe environment. In case the analysis of information gathered by ISTAR predicts a possible terrorist attack, the information can be analyzed and visualized within a virtual environment in order to plan a course of preventive actions; this would improve readiness. Here simulation and 3D visualization might also be useful, for instance, for producing and analyzing different courses of military actions, simulation, and predictive analysis of terrorist attacks in order to prevent new ones. Figure 10.5 shows a 3D representation of Barcelona's tube, where different scenarios and situations can be generated to train personnel. For instance, a trainee might have to learn how to look for an abandoned bag or analyze the scenario where that bag contains explosives, and an evacuation alarm must be produced, and security personnel must ask passengers to abandon the tube station (BBC, 2017). That is, many aspects of a threatening situation can be represented, simulated, visualized, navigated, and analyzed for a variety of purposes, which include prevention, training, planning, and rehearsal, and so on.

Only some examples of virtualytics, both real and under research, have been presented. However, many other possible applications might be mapped from the plane (analytics, VR) into the military dimension as shown in Figure 10.3, even some which today seem to fall into the science-fiction domain. For instance, within military training, we can mention the following.

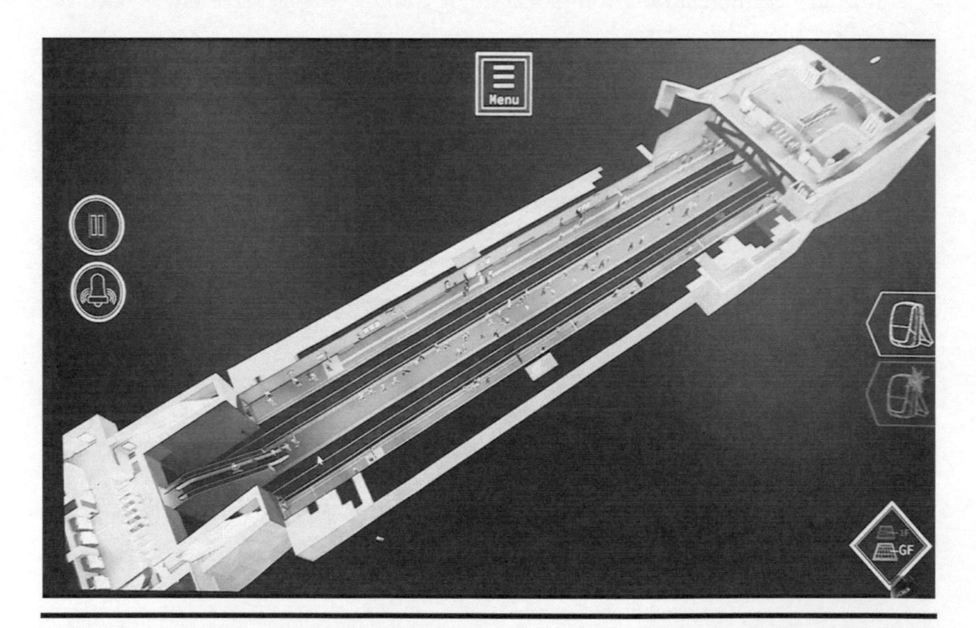

Figure 10.5 3D representation of Barcelona's tube used to train security personnel to act in a possible terrorist attack. (BBC, 2017.)

- Training to deactivate mines or explosives either by humans or by using remotely operated vehicles (ROVs)
- Training to deactivate nuclear bombs
- Training to help/assist people during disasters such as floods, earthquakes, volcanic eruption, and others
- Training to fight against robots/unmanned vehicles
- Training for military activities such as conflict solving, tactics, terrorism prevention, etc.

VR allows for the creation of both real and nonexistent virtual worlds, thus it could be used to train practically any possible real-world military activity such as conflict solving, tactics, and terrorism prevention, to name a few.

Conclusions

Integrating the technologies of analytics and VR into the military field looks promising. The military is an engine that promotes the development of new technologies, as has been the case with VR. The use of analytics in the military is just a natural trend by itself and in combination with other technologies such as VR. Data visualization is only one of the possible contributions to analytics.

By using a 3D cube, some examples of applications into the military were provided, where the main features and benefits of the two dimensions, namely VR and analytics, were mapped into military activities. This cube is open and could be used to map further benefits and features of VR and analytics within the military, and also in other fields, so that other useful applications can be identified.

This cube might be helpful to identify prevention applications, for instance, the recent increase in terrorist attacks (e.g., France, Barcelona, Nevada, etc.), violence in Mexico, and event disasters such as earthquakes and tsunamis, and so on. These might be recreated by building interactive 3D representations and integrating simulations of different aspects involved such as crowd simulation, strategies used by criminals, and water flows so that each situation can be analyzed. In the case of terrorism, what are the main features chosen by terrorists? Do these attacks have common features? Data analysis might throw insights such as crowded places or cities, crowded events, some common failures in safety measures, vulnerability, the attacker's modus operandi, attacker's profiles, and so on. In case of violence in Mexico, analysis might uncover and locate the geographical position of criminal activities filtered at all levels in the country and the lack of operations transparency, pinpointing corruption spots, preference for criminal activities due to extremely low wages and poverty, exhibition of corrupt transnational contracts, and so on. In case of disasters such as earthquakes, crowd simulations, and 3D representations might be helpful to identify optimal exits routes. Flood simulations might be helpful to identify high-risk locations.

In all cases, data analysis might be helpful to prevent human loses or physical and economic harm and to design preventive strategies so that readiness and effectiveness can be increased.

Acknowledgments

All the authors want to thank the valuable comments of reviewers to the previous version of this chapter.

References

Aldin. (2016). *Ghostline, A VR Analytic Suite*. Retrieved from Ghostline: http://ghostline.xyz/

Azuma, R. T. (1997). A survey of augmented reality. *Presence*, 6(4), 355–385.

Baumann, J. (2010). *Military Applications of Virtual Reality*. Retrieved from Human Interface Technology Laboratory: http://www.hitl.washington.edu/projects/knowledge_base/virtual-worlds/EVE/II.G.Military.html

BBC. (2017, June 29). *Can Virtual Reality Train People to Deal with Terrorism?* Retrieved from: https://www.youtube.com/watch?v=DlDZWFXtYmU

Bertolucci, J. (2013). *Big Data Analytics: Descriptive vs. Predictive vs. Prescriptive*. Retrieved from InformationWeek: http://www.informationweek.com/big-data/big-data-analytics/big-data-analytics-descriptive-vs-predictive-vs-prescriptive/d/d-id/1113279

Biehn, N. (2013, May 7). *The Missing V's in Big Data: Viability and Value*. Retrieved from wired: https://www.wired.com/insights/2013/05/the-missing-vs-in-big-data-viability-and-value/

Burdea, G., and Coiffet, P. (2003). *Virtual Reality Technology* (2nd ed.). New Brunswick, NJ: Wiley-IEEE Press.

Cameron, J. (Direction). (1984). *The Terminator* [Película]. United States.

Çintiriz, H., Buhur, M. N., and Şensoy, E. (2015). Military implications of big data. *International Conference on Military and Security Studies*, Istanbul, Turkey.

Craig, A. B., Sherman, W. R., and Will, J. D. (2009). Developing virtual reality applications. In M. Kaufman, *Foundations of Effective Design*. Amsterdam, the Netherlands: Elsevier.

Dunne, J. R., and McDonald, C. L. (2010). Pulse!!: A model for research and development of virtual-reality learning in military medical education and training. *Military Medicine*, 175(7 Suppl), 7–25.

Elliott, T. (2013, July 9). *Business Analytics & Digital Business*. Retrieved from timoelliott.com: http://timoelliott.com/blog/2013/07/the-datification-of-our-daily-lives.html

Ellis, S. R. (1994, January). What are virtual environments? (C. U. IEEE Computer Society Press Los Alamitos, Ed.) *IEEE Computer Graphics and Applications*, 14(1), 17–22.

Fakete, J.-D. (2013). Visual analytics infrastructures: From data management to exploration. *IEEE Computer*, 46, 22–29.

Fletcher, J. D. (2013). What have we learned about Computer Based Instruction in Military Training. In R. J. Seidel and P. R. Chatelier, *In Virtual Reality, Training's Future?: Perspectives on Virtual Reality and Related Emerging Technologies. Defense Research Series. Vol. 6.* New York: Springer Science Business Media, LLC.

Forces, T. (2016, April 19). *Virtual Reality: The Future of Military Training*. Retrieved from: https://www.youtube.com/watch?v=i6pEbpNM1Q4

Fuchs, H., Livingston, M., Raskar, R., Colucci, D., Keller, K., State, A., Crawford, J. R., Rademacher, P., Drake, S. H., and Meyer, A. (1998). Augmented reality visualization for laparoscopic surgery. *Medical Image Computing and Computer-Assisted Intervention — MICCAI'98, Vol. 1496*, pp. 934–943. Cambridge, MA: Springer-Verlag.

Galetto, M. (2017, June 9). *ND DATA: What Is Big Data Analytics?* From https://www.ngdata.com/what-is-big-data-analytics/

Gittings, H. (2017, March). *Analytics' Role in the New Era of Military Readiness. Qlik Blog.* Retrieved from: http://global.qlik.com/us/blog/posts/heather-gittings/analytics-role-in-the-new-era-of-military-readiness

Haar, R. T. (2005). Virtual reality in the military: Present and future. *3rd Twente Student Conference on IT.* Enschede, the Netherlands: University of Twente, Faculty of Electrical Engineering.

Halo. (2016, July). *Descriptive, Predictive, and Prescriptive Analytics Explained.* Retrieved from Halo Business Intelligence: https://halobi.com/2016/07/descriptive-predictive-and-prescriptive-analytics-explained

Haridas, M. (2015). Redefining military intelligence using big data analytics. *Scholar Warrior*, Autum 2015, 72–78.

Höllerer, T. H., and Feiner, S. K. (2004). Mobile augmented reality. Chapter 9. In H. Karimi and A. Hammad, *Telegeoinformatics: Location-Based Computing and Services.* London, UK: Taylor & Francis Group.

Järvinen, P. (2013). *A Data Model Based Approach for Visual Analytics of Monitoring Data.* Espoo, Finland: Aalto University.

Knerr, B. W., Breaux, R., Goldberg, S. L., and Thurman, R. A. (2002). National defense. In K. M. Stanney, *Handbook of Virtual Environments: Design, Implementation and Applications*, pp. 857–872. Mahwah, NJ: Lawrence Erlbaum Associates.

Krevelen, D. V. (2007, April 18). Augmented reality: Technologies, applications, and limitations. *Enabling Technologies*. doi:10.13140/rg.2.1.1874.7929.

Lele, A. (2013). Virtual reality and its military utility. *Journal of Ambient Intelligence and Humanized Computing*, 4(1), 17–26.

Lewis, M. (2004). *Moneyball: The Art of Winning an Unfair Game.* New York: W. W. Norton.

Logie Lab. (2016, March 7). "T-Rex". Logie T. Rex. Realidad Aumentada. Retrieved from: https://play.google.com/store/apps/details?id=com.logie.promotypes.trex&hl=es

Machover, C., and Tice, S. E. (1994). Virtual reality. *IEEE Computer Graphics & Applications*, 14, 15–16.

McGregor, C., Bonnis, B., Stanfield, B., and Stanfield, M. (2017). Integrating big data analytics, virtual reality, and ARAIG to support resilience assessment and development in tactical training. In N. Dias, S. de Freitas, D. Duque, and N. Rod (Eds.), *2017 IEEE 5th International Conference on Serious Games and Applications for Health (SeGAH)*, pp. 1–7. Perth, Western Australia: IEEE.

Milgram, P., and Kishino, F. (1994, December). A taxonomy of mixed reality visual displays. *IEICE Transactions on Information Systems*, E77-D(12), 1321–1329.

Miller, B. (Direction). (2014). *Moneyball* [Película].

Olshannikova, E., Ometov, A., Koucheryavy, Y., and Olsson, T. (2015). Visualizing big data with augmented and virtual reality: Challenges and research agenda. *Journal of Big Data a Springer Open Journal*, 2, 22.

Osarek, J., Frisch, C., Izdebski, K., Legkov, P., Maschmann, M. C., Gordon, C. I., Scholz, A., Sommerer, F., and Williams, K. (2016). *Virtual Reality Analytics: How VR and AR Change Business Intelligence*. In J. Osarek, Bad Homburg, Germany: A Skilltower Institute Dossier.

Parkin, S. (2015, December 31). How VR is training the perfect soldier. Retrieved from: https://www.wareable.com/vr/how-vr-is-training-the-perfect-soldier-1757

Patterson, B. (2017, January 26). *Business Wire a Berkshire Hthaway Company.* Retrieved from: http://www.businesswire.com/news/home/20170126006150/en/Virtualitics-Launches-Platform-Merge-Artificial-Intelligence-Big

Perez R. M., Zabre B. E., and Islas P. E. (2004). Realidad virtual: Un panorama general. (I. d. Eléctricas, Ed.) *Boletin IIE*, 28(2), 39-44.

Pomerleau, M. (2016, May 16). *Big Obstacles Remain for Big Data Analytics at DOD. GCN Technology, Tools and Tactics for Public Sector IT.* Retrieved from: https://gcn.com/articles/2016/05/16/military-analytics.aspx

Ponsoldt, J. (Direction). (2017). *The Circle* [Película].

Retinad. (2016). *Analytics for Virtual Reality*. Retrieved from Retinad: https://www.retinad.io/

Rizzo, A., Parsons, T., and Lange, B. (2011). Virtual reality goes to war: A brief review of the future of military behavioral healthcare. *Journal of Clinical Psychology Medical Settings*, 18, 176–187.

Rizzo, A., Reger, G., Gahm, G., Difede, J., and Rothbaum, B. O. (2009). Virtual reality exposure therapy for combat-related PTSD. In J. LeDoux, T. Keane, and P. Shiromani (Eds.), *Post-Traumatic Stress Disorder*, pp. 375–399. Totowa, NJ: Humana Press.

Rumbaugh, R. (2017). *Defining Readiness: Background and Issues for Congress.* Washington, DC: Congressional Research Service.

Russom, P. (2011). *Big Data Analytics.* Rento, WA: The Data Warehousing Institute.

Salas, E., Oser, R. L., Cannon-Bowers, J. A., and Daskarolis-Kring, E. (2002). Team training in virtual environments: An event-based approach. In K. M. Stanney, *Handbook of Virtual Environments. Design, Implementation and Applications,* pp. 873–892. Mahwah, NJ: Lawrence Erlbaum Associates.

Schmitt, P. J., Agarwal, N., and Charles, J. (2012). From planes to brains: Parallels between military development of virtual reality environments and virtual neurological surgery. *World Neurosurgery*, 78(3–4), 873–892.

Schroeck, M., Shockley, R., Smart, J., Romero-Morales, D., and Tufano, P. (2012). *Analytics: el uso de big data en el mundo real. Cómo las empresas más innovadoras extraen valor de datos inciertos.* Madrid, España: IBM Global Business Services.

Scott, K. D. (2011). *Joint Operations.* Washington, DC: Joint Publication 3-0.

Soares, S. (2012, June). *Not Your Type? Big Data Matchmaker On Five Data Types You Need To Explore Today.* Retrieved from Dataversity, Data Education for Business and IT Professionals: http://www.dataversity.net/not-your-type-big-data-matchmaker-on-five-data-types-you-need-to-explore-today/?cm_mc_uid=37197575099514611638378&cm_mc_sid_50200000=1462049931

Stimmel, C. L. (2015). *Big Data Analytics Strategies for the Smart Grid.* London, UK: CRC Press.

Tradoc. (1997). Chapter 5: The military decision-making process. In T. Headquarters, *Staff Organization and Operations. Field Manual No. 101-5 OPORD*, pp. 5-1–5-31. Washington, DC: Headquarters Department of the Army.

Underwood, J. (2016, May). *Immersive Data Visualization with Virtual Reality.* Retrieved from jenunderwood.com: http://www.jenunderwood.com/2016/05/05/immersive-data-visualization-virtual-reality/

Virtual Reality Society. (2016). *Virtual Reality Society*. Retrieved from: http://www.vrs.org.uk/virtual-reality-military

Wikipedia. (2017a, September 19). *Intelligence, Surveillance, Target Acquisition, and Reconnaissance*. Retrieved from: https://en.wikipedia.org/wiki/Intelligence,_surveillance,_target_acquisition,_and_reconnaissance

Wikipedia. (2017b, September 23). *United States Department of Defense*. Retrieved from: https://en.wikipedia.org/wiki/United_States_Department_of_Defense

Wilson, C. (2008). *Avatars, Virtual Reality Technology, and the U.S. Military: Emerging Policy Issues*. Washington, DC: Congressional Research Service.

Wojciechowski, R., Walczak, K., White, M., and Cellary, W. (2004). Building virtual and augmented reality museum exhibitions. In *Proceedings of the Ninth International Conference on 3D Web Technology* (*Web3D '04*), pp. 135–144. Monterey, CA: ACM.

Wu, M. (2013, March 25). Lithium comunity. Retrieved September 21, 2017, from de Big Data Reduction 2: Understanding Predictive Analytics: https://community.lithium.com/t5/Science-of-Social-Blog/Big-Data-Reduction-2-Understanding-Predictive-Analytics/ba-p/79616

Index

Note: Page numbers in **bold** and *italics* refer to tables and figures, respectively.